American Institute of Chemical Engineers

Distillation Topical Conference and Separations Division Conference

Topical Conference at the
2008 AIChE Spring National Meeting

April 6-10, 2008
New Orleans, Louisiana, USA

Printed from e-media with permission by:

Curran Associates, Inc.
57 Morehouse Lane
Red Hook, NY 12571
www.proceedings.com

ISBN: 978-1-60560-214-1

ISBN: 978-1-60560-214-1

American Institute of Chemical Engineers
Proceedings
Three Park Avenue
New York, NY 10016-5991
Phone: 212-591-8100

www.aiche.org

American Institute of Chemical Engineers

Distillation Topical Conference and
Separations Division Conference
2008

TABLE OF CONTENTS

145a Thermosiphon Reboilers

Nicholas F. Urbanski, Heat Transfer Research Inc., 150 Venture Drive, College Station, TX 77845

Learn the fundamentals of thermosiphon mechanisms and how they apply to thermosiphon designs! Apply guidelines to develop accurate performance assessment. This session provides a foundation for understanding the basis for thermal design of typical industrial heat exchangers.

Key Topics: Inlet Piping • Outlet Piping • Horizontal vs. Vertical • Circulating vs. Once-Through • Subcooling • Film Boiling • Instabilities • Column Sump Designs • "War" Stories

Suggested Participants: Engineers who design and evaluate thermosiphons

75a Condenser Design Workshop

Joseph W. Holmes, Heat Transfer Research, Inc., 150 Venture Drive, College Station, TX 77845

Learn the fundamentals of condensation mechanisms and how they apply to condenser designs! Apply guidelines accurate performance assessment. This session provides a foundation for understanding the basis for thermal design of typical industrial heat exchangers.

Key Topics: Film Condensation on Vertical Surfaces • Condensation on Horizontal Tube Bundles • Interfacial Vapor Shear Effects on Condensation Heat Transfer • Condensation Inside Horizontal Tubes • Desuperheating and Subcooling • Condensation with mixtures • Condensation with Enhanced Heat Transfer Surfaces • Two-Phase Pressure Drop • Venting Inerts • Drainage • "War" Stories

Suggested Participants: Engineers who design and evaluate condenser equipment

Low-temperature plasma modification of polyvinylidene fluoride (PVDF) membrane by graft of styrene in gas

Jian Chen, Jiding Li*, Rongbin Qi, Hong Ye, Cuixian Chen

Department of Chemical Engineering, Tsinghua University. Beijing 100084, China

Abstract

In order to control surface pore sizes of polyvinylidene fluoride (PVDF) membranes and their distribution, low temperature plasma-induced grafting modifications of PVDF were studied to prepare hydrophobe membranes. By argon (Ar) treating and subsequent grafting reaction, a hydrophobe monomer, styrene, was introduced onto PVDF membrane. FTIR was employed to characterize the chemical and physical changes of Ar plasma modified membrane; the FTIR results indicate that the graft reaction occurred on PVDF membrane surface. The surface modifications of PVDF membranes were investigated by scanning electron microscope (SEM), atomic force microscopy (AFM) and differential scanning calorimeter (DSC). It was revealed that styrene monomers were successfully grafted onto PVDF membranes. The water permeability and the solute rejection were measured by PVDF membrane modified in different graft condition. With the increasing of graft time, the pores in the modified membranes became smaller and the distribution of pores narrowed, the pure water flux decreased and the solute rejection increased.

Keywords: Ar low-temperature plasma; Surface grafting; Styrene; PVDF membrane

1. Introduction

For the past years, low-temperature plasma techniques have extensively been used to modify surface properties of polymer membranes with the main aim of enhancing their transport selectivity [1, 2]. It is well known that plasma environments contain neutral species, energetic ions, electrons and photons. The interaction of these plasma particles with the polymers can cause chemical and physical modifications to their surfaces, while the bulk properties remain unchanged. [3]. Accordingly, the plasma-induced graft polymerization not only can introduce basic groups onto membrane surfaces but also can decrease the pore size in membranes. Low-temperature plasma techniques have widely been applied to prepare highly selective composite membranes for gas separation [4], reverse osmosis [5], pervaporation [6], microfiltration[7-8], ultrafiltration[9-10], and so on. For some cases, Polyacrylonitrile (PAN) ultrafiltration (UF) membranes modified by plasma treatments and plasma polymerization were investigated by Tran et al [11]. Weibel et al. applied acrylic acid (AA) vapour plasma treatment to modify the surface properties of polyurethane membranes (PU) [12].

Porous polyvinylidene fluoride membrane (PVDF membrane) has received much attention from researchers in various fields since the PVDF membrane has been found its application in chemical engineering, electronics, textile industry, food processing and biochemistry, for its excellence in chemical stability, thermal stability, and toughness [13]. Oehr et al. studied plasma aminofunctionalisation of PVDF microfiltration membranes and investigated the plasma

*Corresponding author. Tel. +86 10 62782432; Fax: +86 10 62770304
E-mail: lijiding@mail.tsinghua.edu.cn

modifications with a grafting method using ESCA and an amino-selective fluorescent probe [14]. Grafting of aminosiloxane groups on PVDF supports performed in a low-frequency plasma deposition process using a tetramethylsilane/ammoniac mixture was investigated by Roualdes et al [15]. Wang et al. conducted a contrastive research on the surface molecular structure of porous PVDF membrane before and after modification [16]. Mariana et al. used rf Ar plasma to modify the Surface of polyvinylidene fluoride (PVDF) membrane [17].

Our previous works [18-21] investigated the preparations of NF membranes from PAN UF membrane by Ar low-temperature plasma treatment and subsequent grafting in organic vapor. Some typical plasma irradiation conditions such as plasma power of 30-40W, irradiation time of 60s and pressure of 15-20Pa were experimentally selected [18-19]. In this study, in order to increase the hydrophobic property of PVDF membrane and control surface pore sizes of polyvinylidene fluoride (PVDF) membranes and their distribution, the PVDF membrane was treated by argon (Ar) low temperature and subsequent grafted by a hydrophobic monomer, styrene. The aim of this work was the investigation of the influence of plasma condition on the PVDF surface and typical plasma processes.

2. Experimental

2.1 Materials

PVDF (density 1.76g/cm^3, molecular weight 170 000) was purchased from SOLVAY company (Belgium). Triethyl phosphate (TEP) was obtained from Beijing Yili Fine Chemicals Co. Ltd., Beijing, PR China. The other solvents such as styrene and acetone, received from Beijing Fine Chemicals Company, Ltd, PR China. All chemicals used in the experiments were of analytical grade and were used without any further purification. Argon gas, received from Huayuan Gas Chemical Company (Beijing) was of ultra-high purity.

2.2 Membrane preparation

2.2.1PVDF UF membrane preparetion

15wt% of PVDF was dissolved in TEP at 80℃ with stirring. To remove air bubbles, the homogeneous casting solution was kept at room temperature for 24h under vacuum. After degassed, the casting solution was cast on a polyester non-woven fabric with a doctor knife having 150μm thickness. The nascent membrane was dried for 10s at 25℃. And then, it was immersed into DI water at 25℃. After the immersion, the precipitated membranes were washed for 12h to remove residues of solvent mixtures from the membranes.

2.2.2 Plasma-induced graft polymerization

Plasma reactor was pictured in Fig.1 consisting of the following components: a tubular reaction chamber (diameter 80 mm, length 300 mm), a monomer reservoirs, and connecting tube made from glass; a radio frequency (RF) power supply SY300 with matching network (SP-II, 13.56 MHz, made by Microelectronics R&D Centre of Chinese Academy of Sciences) connected with a copper coil surrounding the reaction chamber; a pressure gauge with power supply (type ZDF-III; Chengdu Chenghua Electronic Instrument Factory); Argon tank; and a vacuum pump (Beijing Instrument Factory).

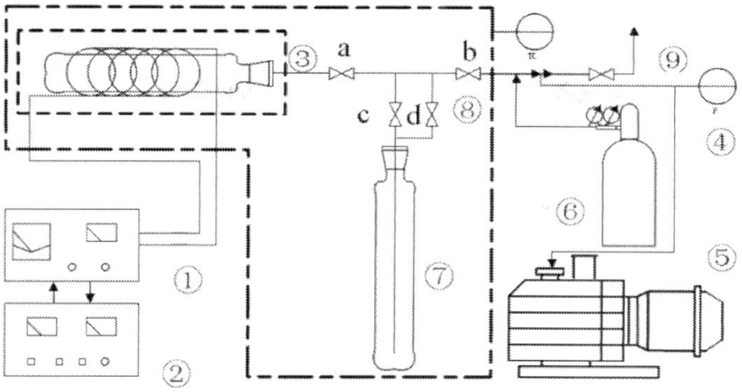

Fig.1 Scheme of experimental equipment for plasma graft

(1)RF power supply, (2) matching network, (3) reaction chamber and copper coil, (4) pressure gauge, (5) vacuum pump, (6) Argon tank (7) monomer reservoirs, (8) glass plunger valve (a, b,c and d) and (9) exhaust valve.

2.3 Membrane characterization

2.3.1 SEM analysis

The surface of virgin PVDF membranes and plasma graft modified samples were taken by scanning electronic microscopy (SEM) (JSM7401 Scanning Electron Microscope). PVDF membranes were coated with Au/Pd three minutes in vacuum.

2.3.2 FTIR-ATR analysis

FTIR-ATR (Fourier transform infrared spectra) spectra were recorded using a Nicolet IR 560 spectrometer with horizontal ATR accessory equipped with a ZnSe crystal. For evaluation, 32 scans were taken with 4cm^{-1} resolution. FTIR spectra were recorded within the range of 4000-400 cm^{-1}.

2.3.3 Atomic force microscopy (AFM) analysis

Atomic force microscopy was a characterization technique, which presented very high possibilities of application in the field of microscopy observation and characterization of various surfaces. The surface morphology of membranes was characterized with atomic force microscope (AFM, SII SPA-400, Seiko).

2.3.4 Pore size determination

The freezing temperature of water held in porous membrane was determined by DSC measurements, i.e., thermoporometry. The DSC apparatus was a Seiko DSC 6200 differential scanning calorimeter equipped with a liquid nitrogen-cooling accessory and calibrated with indium. The sample of about 10-20 mg was put in an aluminum pan and one drop of DI water was added to maintain the sample in an excess of water. Thereafter, the pan was sealed and cooled down to -15℃ at a rate of 1℃/min, which is slow enough to maintain the thermodynamic equilibrium. From DSC thermograms and related equations [22], the pore radius distribution curves were obtained.

2.3.5 Contact angle

The measurement of static contact angle of pure water on membrane surface was made by the liquid-drop method using a goniometric method with a Data Physics OCA20 surface tension meter at room temperature. All grafted PVDF membranes were stored for approximately 5 days after prepared at room temperature in air prior to their contact angle analysis. This was done because their contact angles may initially change after irradiation due to reactions of the irradiated polymers with water vapor and oxygen [23, 24].

2.3.6 Pure water permeation experiments and separation experiments

The pure water flux tests, as shown in Fig.2, were conducted with a cross-flow membrane module, which can offer an active membrane area of 22.4 cm^2 and was made of stainless steel that can endure a pressure difference of 1.0 MPa. DI water was used to measure the pure water flux of the membrane. The prepared membranes were cut into desired size needed for fixing it up in the ultrafiltration kit and initially pressurized with DI water at 0.5 MPa for 6h. The flux generally declines initially and attained steady state after 4-5h of compaction in order to stabilize membranes. Then the water flux was repeatedly measured by collection of the filtrate for given periods until a stable value of the permeation flux (J_w) was observed at 0.1MPa and 25℃. Water flux was calculated as the following equation (1):

$$J_w = \frac{W}{A \cdot \Delta t} \qquad (1)$$

where W is the total volume of the permeate penetrating through the membrane during the experiment, A is the membrane area, and Δt is the operation time.

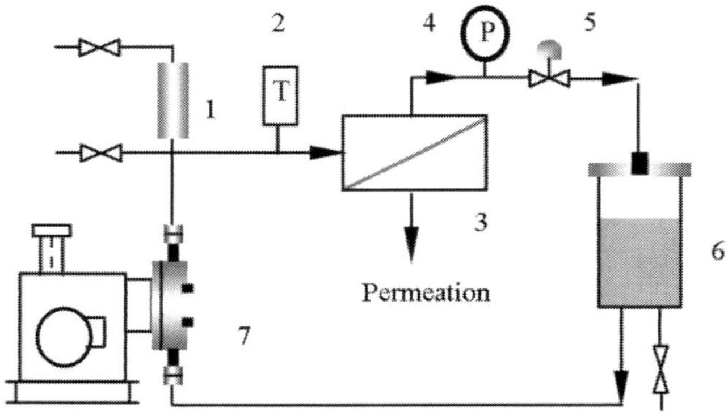

Fig.2 Schematic flow diagram of membrane evaluating equipment

(1)Cushioning,(2) Electric thermometer,(3) Membrane model,(4) Pressure gauge,(5) Valve of adjusting pressure,(6) Feed tank,(7) Controlled volume pump

The solute rejection rate was measured with the bovine serum albumin (BSA) (average molecular weight 68,000g/mol) as the retentate. Values of two membranes were averaged. The solute rejection (R) was determined using equation (2):

$$R = 1 - \frac{C_p}{C_f} \qquad (2)$$

Where C_p and C_f are the retentate concentration in the permeate and in the feed, respectively.

3. Results and Discussion

3.1 SEM photographs of PVDF membrane

The differences in surface morphology between the virgin PVDF membrane and plasma-grafted ones were shown in Fig.3. Virgin membrane surface (Fig.3a) is quite uniform. The surface morphology altered remarkably with the increase of graft reaction time by grafting more and more styrene. It can also be observed from Fig.3 (a) that the surface of PVDF membrane without modification had a great deal of holes as it was ultrafiltration membrane, and the surface changed smoothly as it covered with the styrene with the increase of graft reaction time. After 180 min the membrane surface became more accidented (Fig. 3d) as too much styrene covered the surface of the membrane and then graft reaction occurred. Pure water permeation was mainly dependent on the top layer of the membranes. This probably explained for the low pure water permeation.

Fig.3 Results of SEM of the membranes

(Plasma conditions were Ar plasma, 15 Pa, 40Wand 60 s; Temperature of styrene monomer was 20℃; Grafting reaction times were 0min (a), 30 min (b), 60 min (c), 180 min (d), respectively.)

3.2 FTIR photographs of PVDF membranes

In order to obtain detailed information about the structural changes of PVDF membrane resulting from plasma graft modification, FTIR spectra of the surface of virgin membrane and modified ones were recorded in Fig.4. Compared to the virgin PVDF UF membrane (Fig.4a), the IR spectra (from Fig.4b to Fig.4c) of the membranes grafted with styrene vapor displayed three new absorbance signals at 3022 cm^{-1} (aromatic C-H stretching), 1493 cm^{-1} (C=C stretching) and 695 cm^{-1} (aromatic C-H out-of-plane bending), respectively. These three new peaks are the evidences of the presence of grafted styrene. Furthermore, with the increase in grafting reaction time, IR spectra of the grafted membranes exhibited increased absorbance signals at characteristic band. It seemed that the intensity of these three peaks, especially at 695 cm^{-1}, was correlated to the grafting degree.

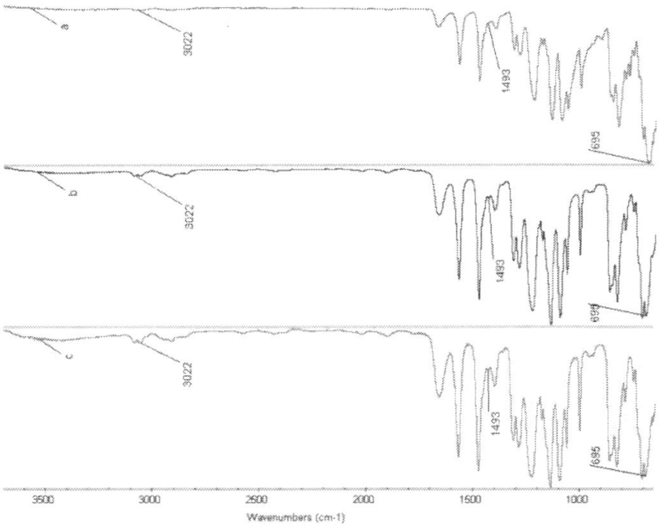

Fig. 4 The FTIR spectra of PVDF modified membrane of different graft time

(Plasma conditions were Ar plasma, 15 Pa, 40Wand 60 s; Temperature of styrene monomer was 20℃; Grafting reaction times were 0min (a), 30 min (b), 60 min (c), respectively.)

3.3 AFM spectroscopy of PVDF membranes

To have a better understanding about the changes in the surface chemistry of membranes, virgin membrane and modified ones were further ascertained by AFM analysis showed in Fig.5. It clearly showed that with the increase of graft reaction time by grafting more and more styrene, the surface morphology altered remarkably. The results indicated that the nodules were seen as bright high peaks whereas the pores are seen as dark depressions. Apparently it can be observed that the PVDF UF membrane without modification contained large numbers of pores, and the membranes surface became rougher after 180min graft reaction time as too much styrene covered. These results were consistent with the results of SEM of the membranes.

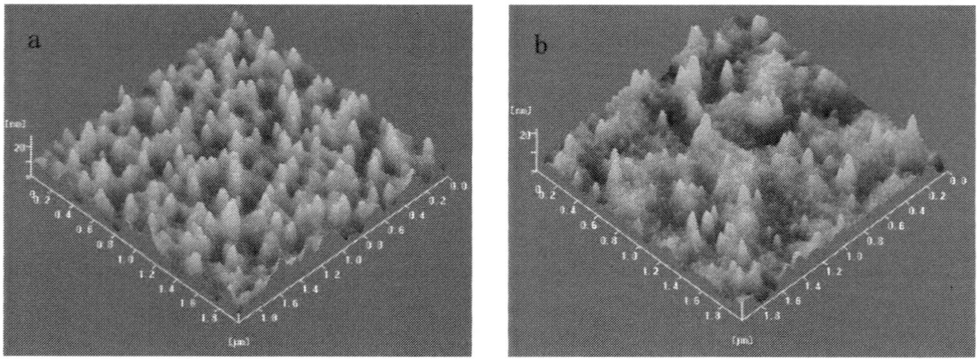

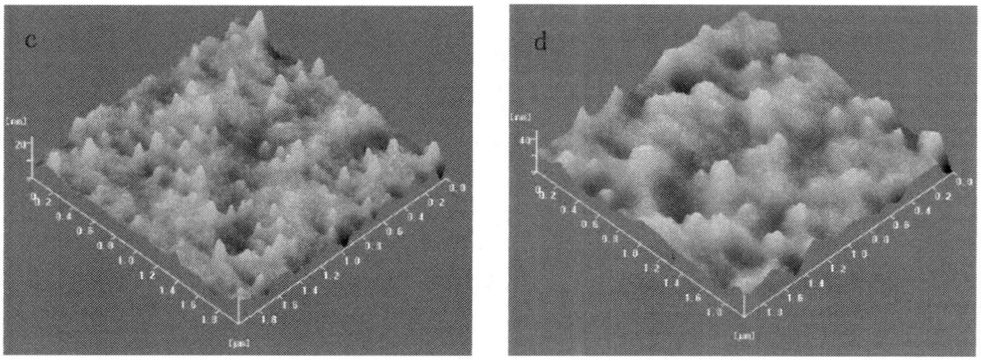

Fig.5 Results of AFM of the membranes

(Plasma conditions were Ar plasma, 15 Pa, 40Wand 60 s; Temperature of styrene monomer was 20℃; Grafting reaction times were 0min (a), 30 min (b), 60 min (c), 180 min (d), respectively.)

3.4 Pore size distribution

Fig.6 exhibited the effects of graft reaction time on pore size and its distribution. From Fig.6a, virgin PVDF UF membrane had pore radius ranging from 20-22 nm. It can be noticed that the average pore radius of membranes became smaller and smaller with the increasing graft time, and its distribution became narrower. Furthermore, increasing graft time causes a decrease of the average pore size from 20-22nm to 13.5-13.8nm and a significant narrowing in the pore size distribution. In this investigation, preparation of hydrophobic nanofiltration membranes has been successfully accomplished using a procedure consisting of plasma irradiation and subsequent grafting in gas of styrene monomer.

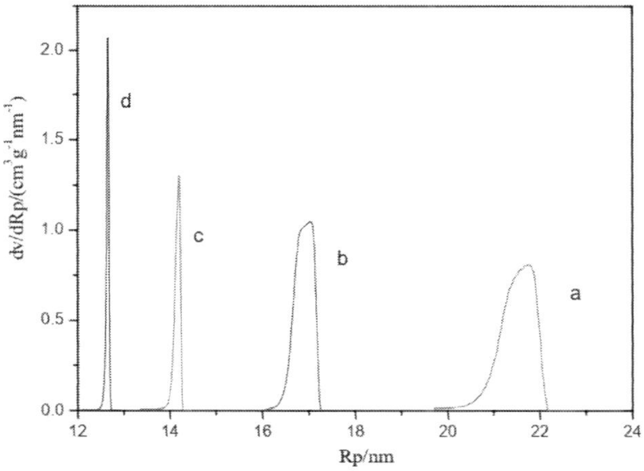

Fig. 6 DSC result of PVDF grafted membrane

(Plasma conditions were Ar plasma, 15 Pa, 40Wand 60 s; Temperature of styrene monomer was 20℃; Grafting reaction times were 0min (a), 30 min (b), 60 min (c), 180 min (d), respectively.)

It should be pointed out that pore sizes determined by DSC technique seem to be larger than the real values [25-26]. But these results from DSC measurement are a valuable tool in understanding plasma graft actions and controlling the membrane structure. The investigation on accurate measurement about membrane pore sizes is in progress.

3.5 Contact angle

Fig.7 revealed the effect of grafting reaction time on contact angle (with standard deviation) of membranes. From Fig.7, the contact angle was increasing with the grafting reacting time. The result suggested that styrene was grafted onto PVDF UF membrane surface and then the hydrophobicity of membrane surfaces increased. It could be educed that, in order to obtain a desired hydrophobic membrane, a feasible way to control the degree of grafting was to regulate plasma conditions, such as plasma power and irradiation time [27].

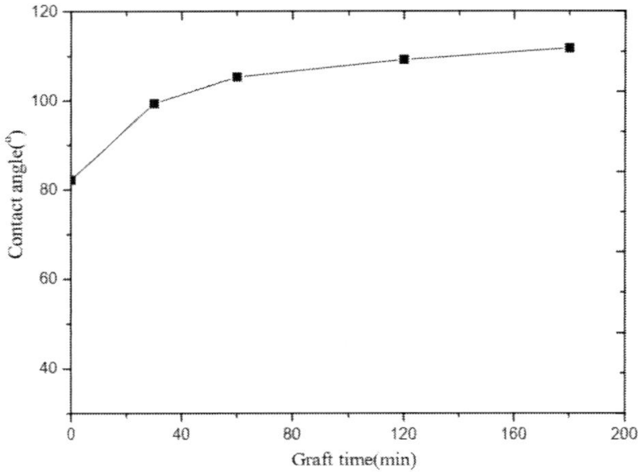

Fig. 7 The effect of grafting reaction time on contact angle (with standard deviation) of PVDF membranes

3.6 Pure water flux experiment and separation experiments

Fig.8 showed the effect of graft reaction time on pure water permeability and solute rejection of PVDF UF membrane. The water permeation flux decreased and thereof the solute rejection increased with the increasing in graft reaction time. This result can be understood in the light of excitation that takes place in plasma treatment. Increase of graft reaction time can excite more active styrene grafted into the surface of the membranes. It was obvious that short graft reaction periods changed the water flux very much and so a sharp decrease was observed at the beginning of 60minc. Furthermore, the flux and rejection alter remarkably within 40min of grafting time. The main reason was that there were more active points available to graft and higher graft ratio at the beginning of reaction. After about 60 min, however, there was no significant decline. Then prolonged graft caused a decline of flux. During the graft process, two changes on membrane surface contributed to this permeation trend. One change was that the grafted styrene caused decrease in pore size. The other change was that the graft modification improved the hydrophobicity of membrane surface. The former and latter both decreased the water permeability of membrane. In conclusion, increasing graft time causes decrease of the water flux from 578 to 23 kg/(m^2h) and increase of the solute rejection from 73% to 92%, which average pore size and a significant narrowing in the pore size distribution.

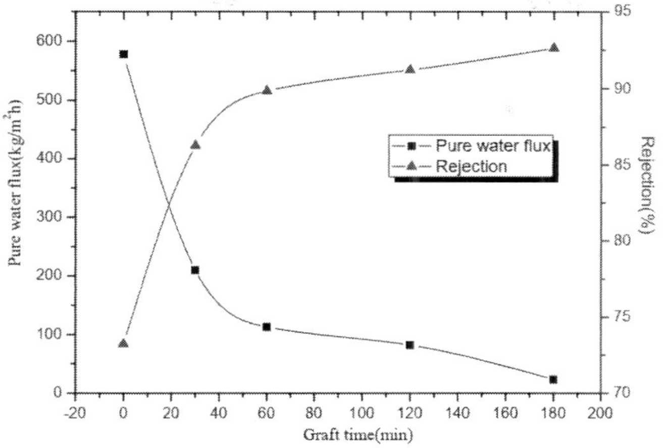

Fig. 8 Influence of graft time on pure water flux of PVDF

4. Conclusion

Plasma is an effective method to change PVDF membrane surfaces' chemical and physical properties. In this investigation, preparation of hydrophobic nanofiltration membranes has been accomplished assembling styrene onto PVDF membrane surfaces using a procedure of plasma irradiation. It was revealed from FTIR spectra, SEM and AFM that hydrophobic monomer, styrene, was successfully grafted onto PVDF UF membrane surface. The contact angle measurement results showed the grafted styrene groups caused a great increase in hydrophobicity of membrane surface grafted sufficiently. Results from DSC measurement demonstrated that the pores in the modified membranes became smaller and the distribution of pores narrowed with the increasing grafting reaction time. Longer graft time caused water flux of PVDF membrane to decrease from 578 to 23 kg/(m^{2}h) at 0.1 MPa and 25℃ and solute rejection to increased from 73% to 92%.

In general, low-temperature plasma technique has been extensively used to modify the polymer membranes as it is now possible to simultaneously adjust hydrophobicity and pore size, hence adjust permeability/selectivity of PVDF UF membranes by plasma-induced graft polymerization.

Acknowledgements

The authors greatly appreciate the financial supports of the Major State Basic Research Program of China (No. 2003CB615701), National Natural Science Foundation of China (No. 20736003， No.20676067, No. 20576059), the National High Technology Research and Development Program of China (863 Program; Grant 2007AA06Z317), SINOPEC Foundation (No. X505002).

References

[1] J. Zhan, Z. Liu, B. Wang, F. Ding, Modification of a membrane surface charge by a low temperature plasma induced grafting reaction and its application to reduce membrane fouling, separation science and technology, 39(2004)13:2977-2995.

[2] P. Favia, L. C. Lopez, E. Sardella E, R. Gristina, M. Nardulli, R. d'Agostino, Low temperature plasma processes for biomedical applications and membrane processing, desalination,199(2006) (1-3):268-270.

[3] Y.M. Lee, J.K. Shim, Preparation of pH/temperature responsive polymer membrane by plasma polymerization and its riboflavin permeation, Polym., 38 (1997) 1227-1332.

[4] K. Li, J. Meichsner, Gas-separating properties of membranes coated by HMDSO plasma polymer, Surf. Coat. Technol. 116-119 (1999) 841-847.

[5] H.I. Kim, S.S. Kim, Fabrication of reverse osmosis membrane via low temperature plasma polymerization, J. Membr. Sci. 190 (2001) 21-33.

[6] M.Y. Teng, K.R. Lee, D.J. Liaw, Y.S. Lin, J.Y. Lai, Plasma deposition of acrylamide onto novel aromatic polyamide membrane for pervaporation, Eur. Polym. J. 36 (2000) 663-672.

[7] D.S. Wavhal, E.R. Fisher, Hydrophilic modification of polyethersulfone membranes by low temperature plasma-induced graft polymerization, J. Membr. Sci. 209 (2002) 255-269.

[8] M.L. Steen, A.C. Jordan, E.R. Fisher, Hydrophilic modification of polymeric membranes by low temperature H2O plasma treatment, J. Membr. Sci. 204 (2002)341-357.

[9] M. Bryjak, I. Gancarz, G. Pozniak, W. Tylus, Modification of polysulfone membranes 4. Ammonia plasma treatment, Eur. Polym. J. 38 (2002)717-726.

[10] K.S. Kim, K.H. Lee, K. Cho, C.E. Park, Surface modification of polysulfone ultrafiltration membrane by oxygen plasma treatment, J. Membr. Sci. 199 (2002)135-145.

[11] T. D. Tran, S.Mori, M. Suzuki, Plasma modification of polyacrylonitrile ultrafiltration membrane, Thin Solid Films, 515 (2007)4148-4152.

[12] D.E. Weibel, C. Vilani, A.C. Habert, C.A. Achete, Surface modification of polyurethane membranes using acrylic acid vapour plasma and its effects on the pervaporation processes, J. of Membr. Sci. 293 (2007) 124-132.

[13] N.A. Ochoa, M. Masuelli, J. Marchese, Effect of hydrophilicity on fouling of an emulsified oil waste water with PVDF/PMMA membranes, J. Membr. Sci. 226 (2003) 203-211.

[14] M. Mu ller, C. Oehr, Plasma aminofunctionalisation of PVDF microfiltration membranes: comparison of the in plasma modifications with a grafting method using ESCA and an amino-selective fluorescent probe, Surface and Coatings Technology, 116-119 (1999) 802-807.

[15] R. Stkphanie, K. Nabila, D. Jean, P. Gerald, Plasma-grafted PVDF polymers as anion exchange membranes for the electrotransport of Cr(V1),Desalination,146 (2002) 273-278.

[16] H. Wang, L. Wang, et al. spectral analysis on the surface molecular structure of modified polyvinylidene fluoride membrane, Applied Surface Science, 252 (2006) 8494-8498.

[17] D. Mariana Duca, L. Carmina, P. Tatiana, surface modifications of polyvinylidene fluoride(PVDF) under rf Ar plasma, Polymer Degradation and Stability, 61(1998) 65-12.

[18] Z.P. Zhao, J. Li, D.X. Zhang, C.X. Chen, Nanofiltration membrane prepared from polyacrylonitrile ultrafiltration membrane by low-temperature plasma: 1.Graft of acrylic acid in gas, J. Membr. Sci. 232 (2004) 1-8.

[19] Z.P. Zhao, J. Li, J. Chen, C.X. Chen, Nanofiltration membrane prepared from polyacrylonitrile ultrafiltration membrane by low-temperature plasma: 2. Grafting of styrene in vapor phase, J. Membr. Sci. 251 (2005) 239-245.

[20] Z.P. Zhao, J. Li, D. Wang, C.X. Chen, Nanofiltration membrane prepared from polyacrylonitrile ultrafiltration membrane by low-temperature plasma: 4. grafting of

N-vinylpyrrolidone in aqueous solution, Desalination 184 (2005) 37-44.
[21] J. Chen, J. Li, Z.P. Zhao, D. Wang, C.X. Chen, Nanofiltration membrane prepared from polyacrylonitrile ultrafiltration membrane by low-temperature plasma: 5. Grafting of styrene invapor phase and its application, Surface & Coatings Technology 201 (2007) 6789-6792.
[22] M. Iza, S. Woerly, C. Danumah, S. Kaliaguine, M. Bousmina, Determination of pore size distribution for mesoporous materials and polymeric gels by means of DSC measurements: thermoporometry, Polym., 41 (2000) 5885-5893.
[23] K. Lunkwitz, W. Bürger, U. Lappan, H.-J. Brink, A. Ferse, Surface modification of fluoropolymers, J. Adhes. Sci. Technol. 9 (1995) 297.
[24] U. Lappan, K.Z. Lunkwitz, Radiation induced functionalization of fluoropolymers, Zeitschrift für Phys. Chem. 191 (2) (1995) 209.
[25] M. Brun, A. Lallemand, J. Quinson, et al. A new method for the simultaneous determination of the size and shape of pores: the thermoporometry. Thermochim Acta, 21(1977)59-88.
[26] Z. Leos, T. Gabriel, Characterization of porous sub-layers in UF membranes by thermoporometry. J. Memb. Sci. 32(1987)329-337.
[27] D.S. Wavhal, E.R. Fisher, Hydrophilic modification of polyethersulfone membranes by low temperature plasma-induced graft polymerization, J. Membr. Sci. 209 (2002) 255-269.

Robust and reliable modelling for a distillation column

Yahya CHETOUANI

Université de Rouen, Département Génie Chimique, Rue Lavoisier, 76821 Mont Saint Aignan Cedex, France. Tel: (0033)235528466, Fax: (0033)235147191, Yahya.Chetouani@univ-rouen.fr

Abstract

In the last few years ever-growing interest has been shown in production quality standards and pollution phenomena in industrial environments. However process development and continuous request for productivity led to an increasing complexity of industrial units. The dynamic nature and the nonlinear behavior of such units pose challenging control system design when products of constant purity are to be recovered. In chemical process control, the processes typically exhibit nonlinear behavior. The intrinsic highly nonlinear behavior in the industrial process, especially when a chemical reaction is used, poses a major problem for the formulation of good predictions and the design of reliable control systems (Chetouani, 2007; Fortuna et al., 2005; Cammarata et al., 2002). Similar problems arise also from the uncertainty for the parameters of the process, such as the reaction rate, activation energy, reaction enthalpy, heat transfer coefficient, and their unpredictable variations. Most large-scale process models derived from first principles are represented by nonlinear differential–algebraic equation (DAE) systems. Since such models are often computationally too expensive for real-time control. Even if global nonlinear models are available, they may not be appropriate for control and monitoring or fault detection purpose (Huang et al., 2000).

Distillation is one of the most important separation processes used in many chemical industries. It can be classified into two ways namely binary distillation for separation of mixture of two substances and multi-component distillation for separation of mixture of more than two substances. Distillation is an energy-intensive separation process in which a liquid or vapor mixture of two or more substances is separated into its component fractions of desired purity by the application and removal of heat. It has extensively been studied (Eden et al., 2000; McAvoy, 1990). There are essentially two approaches by which nonlinear models can be developed for a distillation column; from first principles by using the process knowledge or empirically from input/output data. The advantages and disadvantages of each approach are well known. In industrial practice, it is not always possible in general to obtain accurate first principles models for high-purity distillation columns. Most industrial columns are used to separate multi-component mixtures whose constituent elements are often not known completely; the fundamental thermodynamics of multi-component vapor-liquid equilibrium, the physical property data, and other essential constitutive relations required for the successful development of a first principles model are not always available. And even when such knowledge is available, the resulting models usually occur in the form of a very large system of coupled nonlinear ordinary differential equations, and may therefore not always be the most convenient for controller or fault detection (FD) design. On the other hand, when fundamental process knowledge is unavailable or incomplete, or when the resulting model may not be particularly suitable for FD applications, the input/output models identified from plant data may be more useful. Nonlinear autoregressive models have been used (James et al., 2002), as well as models that use a set of polynomial basis functions (e.g. Volterra functions). Also ANNs provide an excellent mathematical tool for dealing with severe nonlinear problems. Another attractive property is self-learning ability. As a result, nonlinear systems can be modeled with a great flexibility. These features allow one to employ artificial neural networks to model complex, unknown and nonlinear dynamic processes.

The main aim of this paper is to establish a reliable model both for the steady-state and unsteady-state regimes of a nonlinear process. The use of this model should reflect the normal behavior of the process and allow distinguishing it from an abnormal one. In order to obtain this reliable model, the artificial neural network (ANN) is used to model plant input-output data by means of a NARX model. Also this paper shows another technique for neural model reduction into account the physical knowledge of the process. The study shows the choice and the performance of the ANN in the training and test phases. An analysis of the inputs choice, time delay, hidden neurons and their influence on the behavior of the neural estimator is carried out. Three statistical criteria are used for the validation of the experimental data. The model is implemented by training a Multi-Layer Perceptron Artificial Neural Network (MLP-ANN) with input-output experimental data. After describing the system architecture, a realistic and complex application as a distillation column is presented in order to illustrate the proposed ideas concerning the dynamics modelling and model reduction. Satisfactory agreement between identified and experimental data is found and results show that the reduced neural model successfully predicts the evolution of the product composition.

Keywords: Reliability; modelling; artificial neural network; NARX; distillation column; product quality.

References

Chetouani, Y. (2007). Modelling and prediction of the dynamic behavior in a reactor-exchanger using NARMAX neural structure, *Chemical Engineering Communications Journal*, 5, 691-705.

Fortuna, L., Graziani, S., Xibilia, M. G. (2005). Soft sensors for product quality monitoring in debutanizer distillation columns, *Control Engineering Practice*, 4, 499-508.

Cammarata, L., Fichera, A., Pagano, A. (2002). Neural prediction of combustion instability, *Applied Energy*, 72, 513-528.

Huang, Y., Reklaitis, G.V., Venkatasubramanian, V. (2000). Dynamic optimization based fault accommodation, *Computers and Chemical Engineering,* 24, 439–444.

Eden, M.R., Koggersbøl, A., Hallager, L., Jørgensen, S.B. (2000). Dynamics and control during start-up of energy integrated distillation column, *Computer Chemical Engineering*, 24, 1091–1097.

McAvoy, T.J. (1990). Modeling chemical process systems via neural computation, *IEEE Control Systems Magazine*, 3, 24-29.

James, S., Legge, R., Budmann, H. (2002). Comparative study of black-box and hybrid estimation methods in fed-batch fermentation, *Journal of Process Control*, 1, 113–121.

200c Singularities in Non-Equilibrium Reactive Flash Processes

Gerardo Ruiz, Chemical Engineering, UPR Mayaguez, PO Box 9046, UPRM, Mayaguez, PR 00681, Lakshmi Sridhar, Chemical Engineering, University of Puerto Rico, Mayaguez campus, P.O. Box 9046, Mayaguez, PR 00681, and Misael Diaz, Chemical Engineering, Univ of Puerto Rico Mayaguez, PR 00681-9046, MAyaguez, PR 00681.

Multiple steady-states in reactive distillations was demonstrated by several workers (Jacobs and Krishna; 1993, Nijhuis, Kerkhof and Mak, 1993, Bravo, Pylahathi and Jarvelin, 1993, Hauan, Hertzberg and Lien; 1995, 1997; Hauan, Schrans and Lien, 1997; Sneesby, Tade and Smith; 1998a, 1998b; Eldarsi and Douglas, 1998; Mohl Kienle and Gilles, 1998, 1999; Rapmund, Sundmacher and Hoffmann, 1998; Higler, Taylor and Krishna, 1999; Guttinger and Morari (1999a, 1999b); Chen, Huss, Doherty and Malone, 2002). In a recent article Ruiz et al (2006) analyzed the isothermal isobaric reactive flash problem and showed that the MTBE reactive flash process, under isothermal, isobaric conditions exhibited Hopf bifurcations. In this paper we modify the approach used by Ruiz et al (2006), where a continuation procedure implemented in CL_MATCONT was used to solve all the equations for the reactive flash taken together. In this article, we express the reactive flash problem as a set of differential algebraic equation system (DAE) and demonstrate Hopf bifurcations for the MTBE and TAME system even when such an approach is used. We then investigate the nature of the singularities in the non-equilibrium reactive flash problems described in Sridhar et al (2000) and show that the effect of imposing mass transfer rate equations on the equilibrium reactive flash problem causes the occurrence of limit points for the TAME mixture and branch points for the MTBE mixture in single stage non-equilibrium reactive flash problems.

229d Maximize Propylene Recovery from a Fcc Gas Plant

Giuseppe Mosca, Technology, Sulzer Chemtech, Piazza Duca D'Aosta, 12, Milano, 20124, Italy and Norman Burcher, PreemRaff, Lysekil, Sweden.

In the middle of 2002 the FCC Unit at PreemRaff was revamped to boost the capacity and increase the light ends recovery. Several distillatiin columns were retrofitted with advanced Mass Transfer Components, while a new Propylene/Propane Splitter was added at the downstream Gas Concentration Unit. The goal was to maximize the Propylene recovery with a polymer grade quality, which typically requires a huge number of fractionation trays. Due to plant lay-out constraints, there was room for one single column only. To maximize the energy saving a heat pump system was chosen. Due to local Refinery regulations, the column internals had to be designed for an easy accessibility and fast inspection of the vessel; therefore the multiple downcomer trays, typically selected for this duty, could not be used. Following a deep technical study, including process, hydraulic, and mechanical factors, the VGPlus high performance trasy were installed.

The scope of the paper is to describe the approach to such challenging project and report the plant data to highlight the achieved results.

229e Separation of O-Xylene from the Mixed Xylenes by Distillative Freezing

Lie-Ding Shiau and Jung-Hao Chuang. Department of Chemical and Materials Engineering, Chang Gung University, 259 Wen-Hwa 1st Road, Taoyuan, 333, Taiwan

Xylenes have very broad applications in chemical industries. The mixed xylenes produced mainly consist of p-xylene (PX), m-xylene (MX), o-xylene (OX) and ethylbenzene (EB). Due to their similar molecular structures, these isomers have close boiling points. Therefore, it is very difficult to separate them from each other by distillation. The cost for separation and purification of these isomers is generally very high.

A new separation technique, called distillative freezing (DF), has been successfully applied to separate PX from the binary xylene mixture (MX+PX and OX+PX, respectively) (Shiau et al., 2005; 2006). In principle, the DF process is operated at a triple point condition, in which the liquid mixture is simultaneously vaporized and crystallized due to the three-phase equilibrium. Thus, DF is a distillative crystallization technology, which combines distillation and crystallization to produce pure crystals. By lowering temperature and reducing pressure during the operation, the DF process results in the formation of pure crystals, and liquid phase and vapor phase of mixtures. The process is continued until the liquid phase is completely eliminated and only the pure crystals remain in the feed. Distillative freezing (DF) will be applied in this work to separate and produce OX crystals from the mixed xylene mixtures containing MX, PX, OX and EB. The experimental results indicate that, when the DF experiments are performed based on the simulated adiabatic DF operation, OX crystals can be produced and the purity of OX crystals can reach 99%~99.5% for the mixed xylene mixtures of MX, PX, OX and EB.

There are several unique features of the DF operation: (1) The DF process is conducted under an adiabatic condition at a three-phase equilibrium, where the latent heat released in crystallization is removed by vaporization. Thus, it is an energy-conserving separation method compared to distillation. (2) The addition of solvent is not needed in DF compared to azeotropic or extractive distillation in the separation of the close-boiling mixture. Thus, it is a clean separation technique. (3) The DF process is continued until the liquid phase is completely eliminated and only the pure crystals remain in the feed. Compared to crystallization, subsequent filtration or centrifugation is not needed since no mother liquor is present with the pure crystals. (4) Crystal washing is not required since only the pure crystals remain in the feed and no impurities are adhered on the crystal surfaces at the end of the DF operation.

[1]Shiau LD, Wen CC, Lin BS. Separation and purification of p-xylene from the mixture of m-xylene and p-xylene by distillative freezing. Ind. Eng. Chem. Res. 2005;44: 2258-2265.

[2]Shiau LD, Wen CC, Lin BS. Application of distillative freezing in the separation of o-xylene and p-xylene. AIChE Journal. 2006;52:1962-1967.

163b A New Model for Distillation Simulation

Xi-Gang Yuan and K. T. Yu. State Key Laboratory of Chemical Engineering, Tianjin University, 92 Weijin Road, Nankai District, Tianjin, 300072, China

Almost all methodologies for distillation simulation, either equilibrium or non-equilibrium model, are based on the assumption that the liquid and the vapor phases on a tray or at a cross-section of a packed bed are uniformly mixed. As we know that the circulation and channeling are frequently occurred in a column as well as the concentration is also non-uniformly distributed, such assumption may lead to serious error in the simulation. Especially the mal-distributed concentration, it affects highly the mass transfer rate or column efficiency and cannot be ignored. The problem of velocity distribution can be solved by applying the method of computational fluid-dynamics (CFD), while the concentration distribution has not yet been able to predict, but relying on the experimental measurement or using the empirically determined coefficient of back-mixing in the mass transfer equation for approximate estimation. In fact, the accurate prediction of concentration distribution, especially for large distillation column, is necessary for the reliable evaluation of column efficiency, as it is obtained through the summing up of all local mass transfer rates. Therefore the prediction of concentration distribution without depending on experiment or empiricism becomes an important topic need to further investigate.

In this work, a modeling approach is developed for distillation simulation with consideration of both the non-uniform distribution of velocity and concentration. The idea is to incorporate with the CFD methodology for describing the fluid flow, and, in addition, a differential mass transfer equation for predicting the concentration distribution. For closing the differential mass transfer equation, a new model was developed, which consists of the concentration fluctuation variance c^2 and its dissipation rate ε_c. A set of differential equations is then established and can be solved by using the commercial software, like FLUENT. The proposed model, which may be called computational mass transfer, has been applied to predict the efficiencies of the industrial tray and packed columns and found to agree with the reported experimental data.

Acknowledgement: NNSFC's support (grant no 20736005) and PhD students worked on this research should be acknowledged.

93e Standardization of Structured Packing Efficiency Measurements

Zarko Olujic[1], Michael Joedecke[2], and Gunter Schuch[2]. (1) Process & Energy, Delft University of Technology, Leeghwaterstraat 44, Delft, 2628 CA, Netherlands, (2) Process Engineering, BASF Aktiengesellschaft, GCT/C - L 540, Ludwigshafen, 67056, Germany

A major consequence of the ever growing need for bulk-chemicals and the extent of grow of process industries in Far East is that larger and larger equipment is built to stay competitive. The amounts of capital and energy involved require as tight as possible column designs and this posses challenges on scale-up know-how side, which, in turn, relies mainly on efficiency and hydraulics' data obtained from total reflux experiments carried out with much smaller equipment using one of standard test systems, often at conditions different from that of intended application.

Although a lot of experimental work has been done in the past and well established large companies have also collected substantial empirical knowledge from operation of their plants, the design of new as well as redesign of existing columns still encompasses a wide range of concerns. These are mainly related to uncertainties with respect to expected hydraulic and mass transfer performance of novel, performance improving gas/vapor-liquid contacting devices, i.e. trays, random packing, and structured packings, developed and brought to the market mainly by equipment manufacturers. The main stream of such developments is concerned with capacity increase (revamps-retrofits), which implies reaching the capacity increase goal without affecting the separation efficiency adversely, or equipment size reduction (new designs), accompanied (desirable) by energy saving per unit of product (reduction of external reflux). The certainty regarding the hydraulic and mass transfer performance of the G-L contactor considered plays a crucial role for success of such a total cost reduction effort. In order to get this certainty, i.e. to minimize risks associated with building very expensive industrial plants, user companies prefer to have results of dedicated pilot plant scale tests, to be used as basis for validation of own predictive models. Regarding the scarcity and discrepancies among available test installations, establishing standardized testing procedures is considered to be a useful measure to arrive at more confidence in this respect.

With this in mind, and wishing to provide a basis for rising the public knowledge in this respect, this presentation focuses on the uncertainties around experimental characterization of performance characteristics of structured packings, more precisely on clearing the issues associated with usability of total reflux measurements - for (re)design purposes - carried out with some of established test systems using pilot- and semi-industrial scale test installations.

Web Page: www.pe.tudelft.nl

Successful Industrial / Academia Collaboration

Mitchell E. Loescher, PhD, PE
CDTECH

Abstract:

Collaborations between academia and industry can yield very fruitful results. Dr. Seibert and his group at the University of Texas at Austin, have provided vital contributions to our business. Other researchers in academia can look to Dr. Seibert and his group as an example of how these relationships can be formed and maintained. This paper highlights some important criteria for collaborative success.

Interdisciplinary Learning for Chemical Engineering Students from Organic Chemistry Synthesis Lab to Reactor Design to Separation

Matt Armstrong, Richard L. Comitz, Andrew Biaglow, Russ Lachance, Joseph Sloop

Abstract

A novel approach to the Chemical Engineering curriculum sequence of electives here at West Point enabled our students to experience a much more realistic design process, which more closely replicated a real world scenario. Students conduct the synthesis in the organic chemistry lab, then conduct computer modeling of the reaction with ChemCad and Mathematica, analyze chemical separation processes, and design a reactor system. This interdisciplinary learning approach demonstrated to students that all of their courses are meant to compliment each other, their learning, and experiences.

Introduction

The Chemical Engineering curriculum at the United States Military Academy has the students enrolled in three electives simultaneously in the Spring semester of their 3rd year. The electives taken simultaneously are Organic Chemistry II, Separation Processes and Chemical Reaction Engineering. This juxtaposition allowed us to simultaneously study a common reaction, the Friedel-Crafts alkylation, in each of the respective classes.

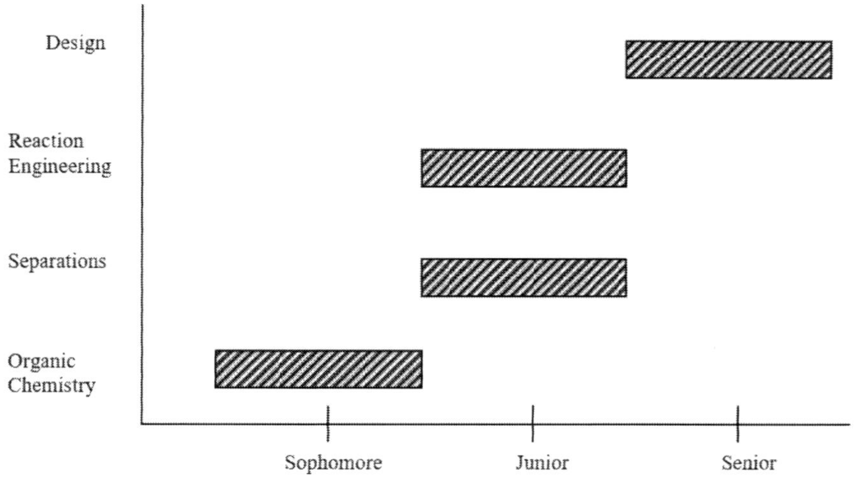

Figure 1. Typical Chemical Engineering Program Order of Electives

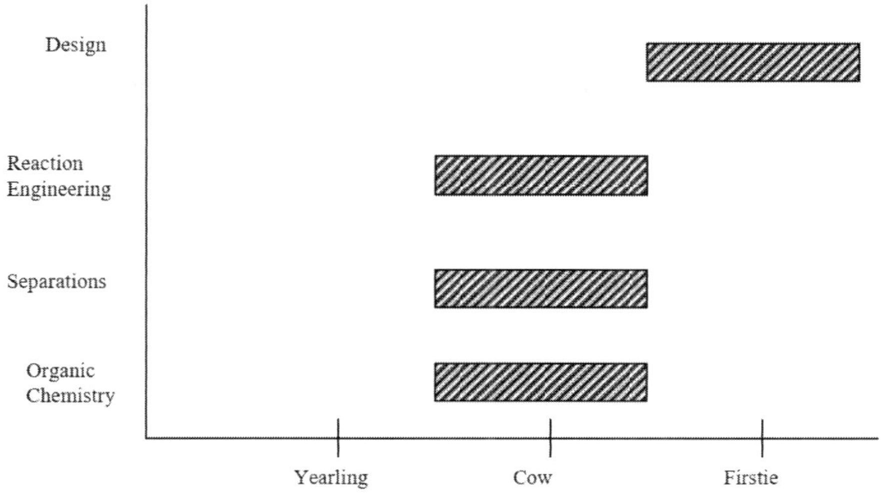

Figure 2. USMA Chemical Engineering Program Order of Electives

The arrangement, therefore, more closely mirrored the real world Chemical Engineering design process. In addition to its realism, the engineering design process used by our students, in addition to its realism also parallels the Military Decision Making Process (see Figure 8), thus reinforcing military as well as engineering decision making concepts.

Lastly our novel approach to curriculum development allowed for an earlier incorporation of the actual data into the process via ChemCad, the Chemical Engineering software that was used by each student. Typically, use of this software does not occur until later in the design sequence.

Background

The Friedel-Crafts reaction is used in laboratory synthesis as well as in industry in the synthesis of ethylbenzene and its derivatives as an intermediate to make styrene monomers[1]. Therefore, this reaction was a good choice to integrate several different courses.

Laboratory experiments conducted during the second semester of organic chemistry generally illustrate practical application of topics covered in lecture. A convenient Friedel-Crafts alkylation reaction which demonstrates the utility of electrophilic aromatic substitution and carbocation rearrangement is that of *p*-xylene with 1-bromopropane yielding approximately a 1:2 ratio of *n*-propyl-*p*-xylene to isopropyl-*p*-xylene.[1] (See Figure 3 and Figure 4).

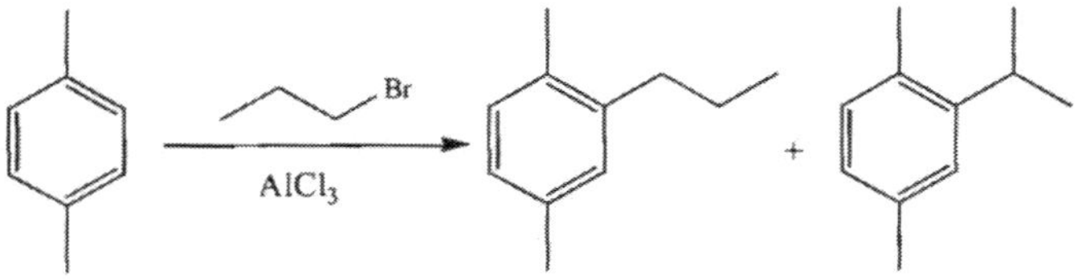

1 : **2**

b.p (°C) **138** **204** **196**

Figure 3. Friedel-Crafts alkylation of *p*-xylene.

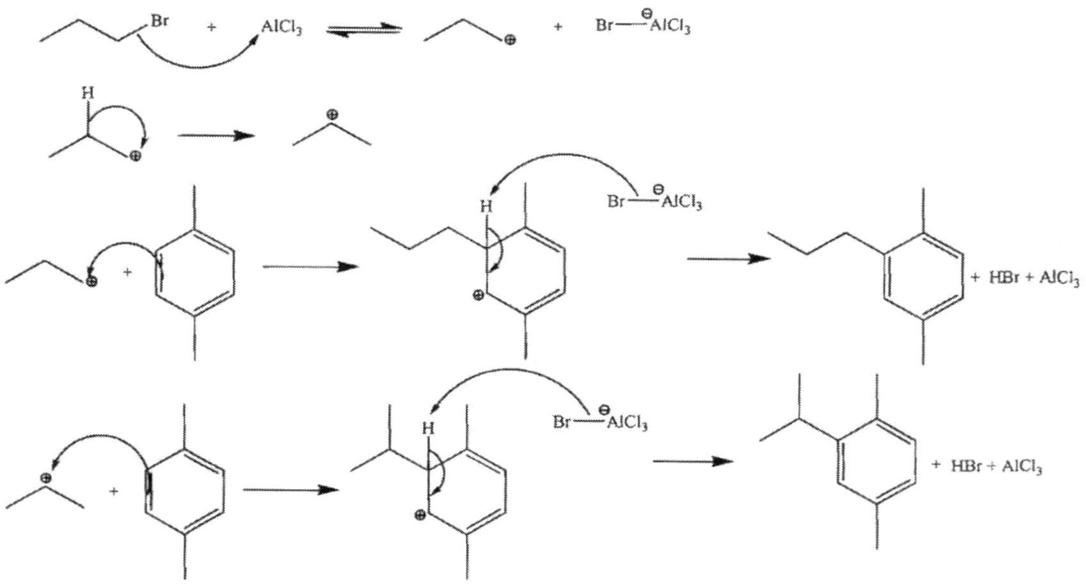

Figure 4. Friedel-Crafts alkylation of *p*-xylene mechanism[1].

Even with activated arene systems like *p*-xylene, carbocation rearrangement leads to a substantial proportion of the isopropyl *p*-xylene. Given that the boiling point difference between the isomeric *p*-xylenes is only 8°C, typical microscale distillation techniques and equipment are not adequate to fractionally separate the isomers. So, although the reaction is satisfactory from a synthetic standpoint, the inability to isolate isomerically pure products leaves students with a problem.

For chemical engineering students, it seems a natural progression to explore solutions to this problem in the context of a chemical separations issue and reactor design. Since these students often take organic chemistry, chemical reactor design, and chemical separations together, an interdisciplinary project such as this provides a practical application to bridge the theory developed in all three courses with an experimental challenge. With our sequencing of electives we have provided our students with a more realistic approach that more closely resembles the reality of the actual design process, to include the ability to use Chemical Engineering software in an earlier stage of the development process.

Results and Discussion

Chemical Reaction Engineering Design Project

In the Chemical Reaction Engineering class, the students were given a design project with the following specifications: 1. Volumetric flow rate υ_0 is 52 L/min; 2. A desired product ratio of 50:50 n-propyl-p-xylene to isopropyl-p-xylene at the outlet; and 3. T_{min} is 15°C and T_{max} is 70°C. The students were directed to use ChemCad to develop their designs, but ChemCad needs frequency factor and activation energy values to correctly model the reactions mathematically. Since these values could not be found in the literature, it was necessary to conduct some preliminary experiments to gather data that the students could use to calculate the frequency factor, k_0, and activation energy, E_a, of each parallel reaction, and the overall reaction. Three independent experiments were run at different temperatures to collect the data required for the concentration vs. time plot. These plots were then used to find reaction rate constants, k, for each temperature for each parallel reaction. The kinetic data was collected following the same procedures the students used in the organic chemistry laboratory earlier in the semester.

To calculate the total reaction rate constant a plot of $C_{bromopropane}/C_{p\text{-}xylene}$ vs. time was constructed. To understand this leap it is necessary to derive the irreversible bimolecular-type second order reaction[2] performance equation:

Starting with the generic second order reaction:

$$\mathbf{A + B \rightarrow products} \qquad (1.1)$$

The corresponding rate equation is as follows[2]:

$$-r_A = -\frac{dC_A}{dt} = -\frac{dC_B}{dt} = k_{tot} C_A C_B \qquad (1.2)$$

It is possible to follow the derivation of this equation in <u>Chemical Reaction Engineering</u>, by Octave Levenspiel in Chapter 3. The following is the end result of the derivations:

$$ln\frac{1-X_B}{1-X_A} = ln\frac{M-X_A}{M(1-X_A)} = (C_{B0} - C_{A0})k_{tot}t \qquad (1.3)$$

where $M = C_{Bo}/C_{Ao}$.

The implication of this result show that a plot of ln (C_B/C_A) versus time will yield a straight line if indeed the reaction is second order, and first order with respect to each reactant. The intercept will equal M, and the slope will be equal to $(C_{B0}-C_{A0})k_{tot}$.

The reaction progress was monitored by gas chromatography, and the kinetic data recorded in Table 1. By plotting the concentration data from the gas chromatograph found in Table 1, it is possible to calculate the $k_{tot.}$[2]

Temp (K)	Time (min)	[Xylene] (*M*)	[$CH_3CH_2CH_2Br$] (*M*)	C_B/C_A	ln (C_B/C_A)
295.5	0	5.29	3.859	0.729489603	-0.31541016
	10	3.79	2.36	0.622691293	-0.4737044
	14	3.63	2.2	0.606060606	-0.50077529
	18	3.34	1.91	0.571856287	-0.55886756
	20	3.18	1.75	0.550314465	-0.59726541
311	0	5.29	3.859	0.729489603	-0.315410163
	2	3.8	2.369	0.623421053	-0.472533142
	6	3.3	1.97	0.596969697	-0.515888926
	14	3.1	1.6	0.516129032	-0.661398482
	18	2.88	1.42	0.493055556	-0.707133423
333	0	5.29	3.85	0.72778828	-0.317745
	2	2.83	1.4	0.494699647	-0.703804
	6	2.39	0.96	0.40167364	-0.912115
	10	2.04	0.61	0.299019608	-1.207246
	14	1.73	0.3	0.173410405	-1.752094

Table 1. Friedel-Crafts alkylation kinetic data

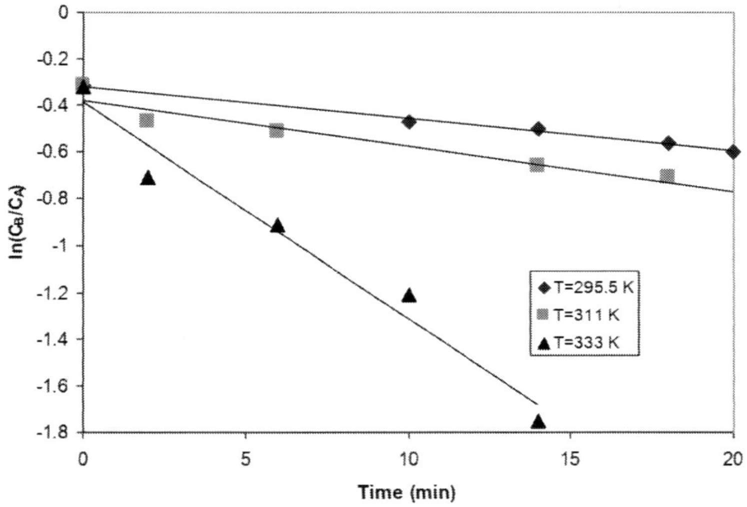

Figure 5. Concentration versus time plot.

With that information and the average ratio of products at each time step it is possible to calculate k_1 and k_2 with the following two equations[2]:

$$k_{tot} = k_1 + k_2 \tag{1.7}$$

$$\left(\frac{C_{n-propyl}}{C_{isopropyl}}\right)_{AVE} = \frac{k_1}{k_2} \tag{1.8}$$

When all of the reaction rate constants were determined it was then possible to solve for individual frequency factors, k_0, and activation energies, E_a, using the Arrhenius relationship:

$$k = k_0 e^{-E_a/RT} \tag{1.9}$$

Plotting ln k vs 1/T, the slope of this line is $-E_A$, and the y intercept is k_0, thus permitting the calculation of both k_0 and E_a for each parallel reaction, and the overall reaction.

This information is critical to model and scale up the reaction using ChemCad. This entire process was expected to be executed by each student, thus reinforcing the derivation of a concentration versus time model. Each student had to demonstrate mastery of this process before using ChemCad at a desk side briefing to the instructor. Upon successful calculation of the reaction rate constants, students were allowed to start the scale up modeling with ChemCad.

With this data, it was now possible to establish the appropriate kinetic relationships in ChemCAD. The students then used ChemCad to search the most economically feasible reactor design. A cursory analysis of the data yielded an appropriate plot of $1/-r_A$ vs. X_A. Analysis of the plot makes it clear that the best reactor design to minimize volume should be a plug flow setup. Using Mathematica, the mean residence time and volume for the initial guess can be estimated. Questions left to resolve are reactor volume, heat duty, and isothermal versus adiabatic operation. Students were free to explore various reactor networks, such as parallel versus series reactors and use of recycle. Students were given latitude to explore other unique strategies using ChemCad.

Chemical Separations Design Project

The chemical separations design phase of this interdisciplinary project was fairly open-ended. The students could use any combination of separations schemes to achieve 90% purity of all components in the system (feed, catalyst, products) and then attempt to achieve a 95% *n*-propyl-*p*-xylene product stream. This open-ended approach forced the students to consider all aspects of a realistic separation problem that originated in their organic chemistry lab and that they might see in industry. At first, the students were intimidated because a detailed solution required knowledge beyond their current level, but they eventually enjoyed working on this problem because it truly challenged them to think.

Like the reactor design project, our students began the separations design project by gathering property information. When they could not find certain property information for some of the compounds they quickly learned how to make reasonable approximations and assumptions. We advised the students that a critical task in their design was to determine the best separation technique for each of the components and decide on the most logical sequencing of those techniques. Based on the available property information, most student teams chose to flash off HBr, extract $AlCl_3$ using water, and use a series of distillations columns to purify the remaining components. However, much like a real-world design process, we forced each team to consider at least two different separation sequences and compare and contrast them. In this way our students learned a great deal about separations processes.

The separations design project also used ChemCad software as the vehicle for the design. Most student teams attempted to jump right into ChemCad without much preparatory analysis, and their initial results clearly emphasized the importance of choosing a reasonable thermodynamic model, and making some preliminary estimates. While students will be expected to use thermodynamic modeling in greater depth later in their curriculum, this exercise served as an excellent tool to emphasize the importance of material yet to come. As a result of creating, manipulating and running ChemCad examples, all students increased their ChemCad proficiency which is a critical software thread for our entire chemical engineering program.

One capstone design team surprised us by exceeding our expectations for a truly integrated design solution. This team combined their reactor design with their separations design in the same process flow sheet. Although we expected separate reactor and separations designs from these 3rd year students in these separate courses, this team made the logical leap and combined the designs to achieve some additional efficiencies. Figure 6 depicts their ChemCad design flow sheet which incorporates a recycle stream for unconverted reactants.

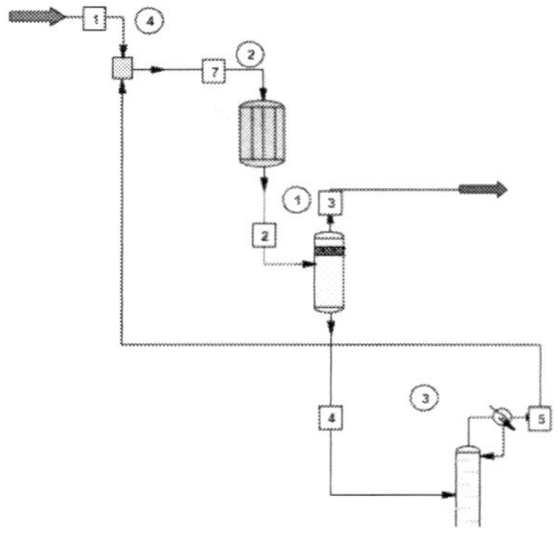

Figure 6. Student Team Fully Integrated Reactor and Separations Design Proposal

Unanticipated benefits

As our students navigated through both projects, we experienced multiple unexpected benefits; some of these have been discussed above. Another significant added benefit was a connection we began to draw between the engineering design process and the Military Decision Making Process (MDMP) taught in 3rd year military science class. As you can see from Figures 7 and 8, both processes first define the problem or the mission by examining facts, assumptions, and specified/implied/critical tasks. Both processes then design alternatives and model or test those alternatives so they can be analyzed and compared. Finally, both processes enable us to arrive at a reasonable decision and both are iterative in nature with feedback loops to further refine the design or plan. While this interdisciplinary project was designed to show our students the connections between organic chemistry, reaction engineering and separations, but we were able to draw multiple connections across many aspects of our curriculum like the case of engineering design and military science.

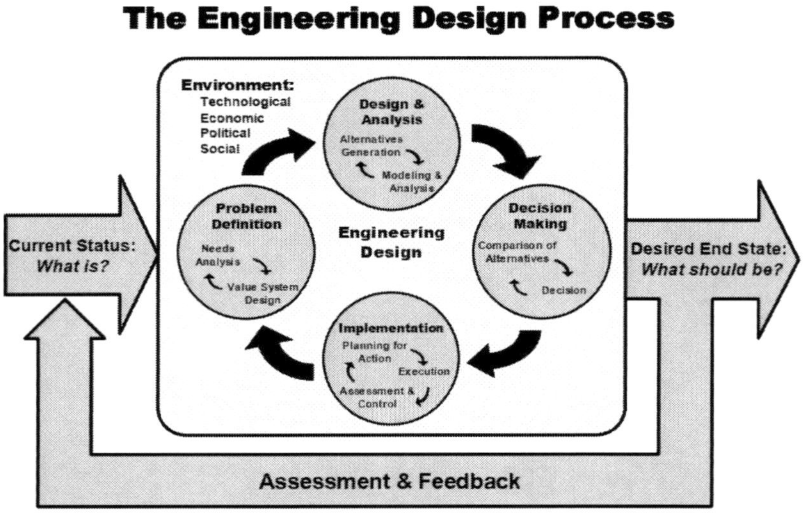

Figure 7: The Engineering Design Process[4]

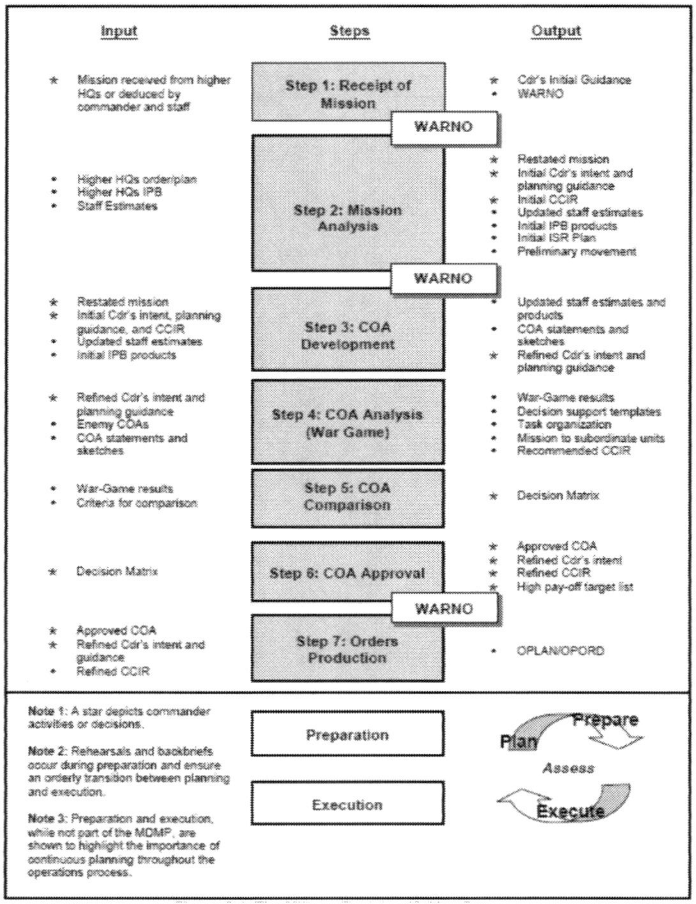

Figure 8. The Military Decision Making Process[3]

Analysis of Results

To analyze the results the students were given a quiz consisting of representative questions from the organic chemistry, chemical reaction engineering and separations disciplines. The same quiz was then re-administered at the end of the project cycle to see if there was improvement, and retention of knowledge. These results are in Table 2.

	Pre-Project:			Post-Project:		
Question	Question #	Correct	Incorrect	Question #	Correct	Incorrect
What is a Friedel Crafts alkylation?	**1**	*5*	*6*	**1**	*7*	*4*
Give an example of one.	**2**	*3*	*8*	**2**	*9*	*2*
Method of calc. k_0 and E_A	**3**	*2*	*9*	**3**	*8*	*3*
Method of k_1, k_2 calc. parallel rxns.	**4**	*0*	*11*	**4**	*7*	*4*
Can k_1, k_2 be found graphically?	**5**	*0*	*11*	**5**	*2*	*9*
Give two ways to separate gas and liquid phases.	**6**	*8*	*3*	**6**	*10*	*1*
Give two ways to separate two liquid phases.	**7**	*8*	*3*	**7**	*10*	*1*

Table 2. Quiz Results

In addition to this the students were asked the following questions regarding their individual experiences with the capstone project. These questions were answered on a scale of 1 to 5, where 1 represented the most positive feedback and 5 was the least positive. These questions are listed below in Table 3 accompanied by the average response. Finally, a comparison will be made of final examination results from AY06-02 to AY07-2 in the chemical reaction engineering course, to see the impact this had on performance.

Question Regarding Individual Experience	Ave Response
1. Was this capstone project useful in terms of helping the learning process?	1.64
2. Was this capstone project helpful to wrap up the course material at end of semester?	1.73
3. Did this capstone project aid your learning in organic chemistry and separations?	2.27
4. Would you recommend this project format next year?	2.09
5. Did you like the capstone project?	2.55
6. Do you think the capstone experience helped your Term End Exam preparation?	2.09

Table 3. Questions Regarding Individual Experiences

From the results, it is clear that the Capstone experience had a positive outcome in terms of mastery of the material. The students' responses to the questions were also quite positive. We will conduct the same approach next year and continue to gather data.

Conclusion

This idea started out as merely a project for our Chemical Reaction Engineering course, but evolved into a novel educational approach to Chemical Engineering curriculum development using a technique closely paralleling the actual industry design process. From our results, it is apparent that this is indeed a valid approach. In fact, we will execute the project again this year to gather more data. The experience allowed the students to approach the problem as a design engineer in industry would, as well as use the problem solving techniques previously discussed. Additionally, to this the students were able to use the Chemical Engineering software earlier by using the kinetic data given to them. We intend to use this technique again, and recommend it fully to other programs.

Experimental

Three experiments were set-up identically at temperatures of 295.5 K, 311 K, and 333 K. To 15.0 mL of *p*-xylene was added 1.00 g of $AlCl_3$. The resulting mixture was allowed to stir while 8.0 mL of 1-bromopropane was added dropwise over a period of 5-10 minutes. At two minute intervals, a microliter sample was extracted from the reaction vessel, quenched with water, and diluted with diethyl ether. After removal of the aqueous layer, the samples were dried over sodium sulfate. The samples were examined in the Gas Chromatograph/MS to determine the concentrations of reactants and products in each sample.

[1] Gilbert, John C. and Stephen F. Martin, Experimental Organic Chemistry A Miniscale and Microscal Approach, *Fourth Edition*, (2006).
[2] Levenspeil, Octave. Chemical Reaction Engineering. *Third Edition,* (1999).
[3] FM 5-0. Army Planning and Orders Production. January 2005.
[4] CE300 Course Book, USMA, 20 June 2007.

BIOGRAPHIES

Colonel Russ Lachance is an Academy Professor supporting our Chemical Engineering Program. His teaching responsibilities include all chemical engineering courses and general chemistry courses. He received his B.S. degree from the United States Military Academy, at West Point NY;his M.S. and Ph.D. in chemical Engineering from MIT, Cambridge, MA. He is the Head Department Academic Counselor for our Department, the ABET coordinator for the Department, and serves on the ABET Committee, the RCI committee and the Faculty Council. He is the Officer-in-Charge of the Cadet Spirit Band, a football mentor, and the PME2 Team Leader for Company B4.

Dr. Andy Biaglow teaches chemistry core and elective courses. He is the director of the Chemical Engineering Program, and responsible for establishing a research program in Chemical Reaction Engineering. Dr. Biaglow received his B.S in Chemical Engineering at Case Western and his Ph.D in Chemical Engineering from University of Pennsylvania. He also did post-doc work at the University of PennsylvaniaDr. Biaglow is a member of the International Zeolite Association, the American Institute of Chemical Engineers, and the American Society of Engineering Educators. He conducts research on reaction pathways and intermediates in solid-catalyzed reactions. He is a member of the Materials Research Society, New York Catalysis Club and the American Chemical Society.

Lieutenant Colonel Joseph Sloop teaches CH383/384, Organic Chemistry I & II. Lieutenant Colonel Sloop's scientific interests include heterocycle synthesis, substituent effects on reactivity in organic systems, and magnetochemistry. He received his B.S. in Chemistry from Davidson College; his M.S. and Ph.D in Chemistry from North Carolina State University, in Raleigh, NC. He is a member of the American Chemical Society, Phi Lambda Upsilon, Phi Kappa Phi, Gamma Sigma Epsilon and the Chemical Corps regiment.

Major Rich Comitz taught CH101/102, General Chemistry last year. He is now the Assistant Course Director for CH383/384, Organic Chemistry I & II, and CH 489/CH490 Individual Research He received his B.S. in Chemistry at the United States Military Academy, at West Point, NY, and his M.S. in Chemistry at the Florida Institute of Technology. MAJ Comitz also serves as an Officer Representative of the Sprint Football Team, and Company Academic Counselor.

Major Matthew Armstrong teaches CH101/102, General Chemistry, CH 364 Chemical Reaction Engineering, CH 489/CH490 Individual Research and also serves as the Department S4. He received his B.S. and M.S. in Chemical Engineering from Rensselaer Polytechnic Institute in Troy, NY. MAJ Armstrong also serves as an Officer Representative for the Cadet Marathon Team, and Company Academic Counselor.

Membrane Contactor/Ionic Liquid for Ethanol Recovery

Lei Sun, Yupo Jim Lin, Mike Henry and Seth W. Snyder
Energy System Division, Argonne National Laboratory

Ethanol is receiving widespread attention in the U.S. as an alternative fuel source. It can be produced by fermentation from a wide variety of biomass feedstocks. However, the fermentation process only yields 2-10% of alcohol. At such low ethanol concentrations, using conventional distillation process to recover ethanol consumes a lot of energy, around 19,000 btu/gal. Majority of the energy is used to remove 90%+ water. Therefore, it is our intention in this work to develop separation technology that can significantly reduce the energy usage for biofuel recovery from fermentations.

Liquid-liquid extraction is a low energy separation technique. Membrane contactor is a membrane-based liquid-liquid extraction process. In this system, pores in the membrane act as a contacting space for two phases. The extracted species is separated from the original phase on one side of membrane into another phase on the other side of the membrane. The membrane provides a physical barrier to prevent emulsion of two phases and also offers a large phase contacting area for liquid-liquid extraction. Since the membrane contactor reported here is used for liquid-liquid extraction, we used ionic liquid (IL) as the extracting solvent for ethanol recovery. ILs are salts that are comprised entirely of cations and anions and have a melting point below the boiling point of water. Like most of the pure salt, IL has near zero vapor pressure. This property makes IL the perfect extraction solvent and attracting a lot of industrial application. They are environmentally benign and for this reason they are considered as "green" solvents. They have been demonstrated for applications in a variety of processes including catalytic reactions, separations, electrochemistry, and biocatalysis. Some of the hydrophobic ILs are completely misicible to ethanol and thus make them very good candidates for ethanol recovery from fermentation broth. We screened several of the ILs which have such properties. Figure on the right shows the results of ethanol separation from water ethanol mixtures using one of the screened ILs--1-hexyl-3-methyl-imidazolium bis(trifluoromethylsulfonyl)imide ([Hmim][Tf_2N]).

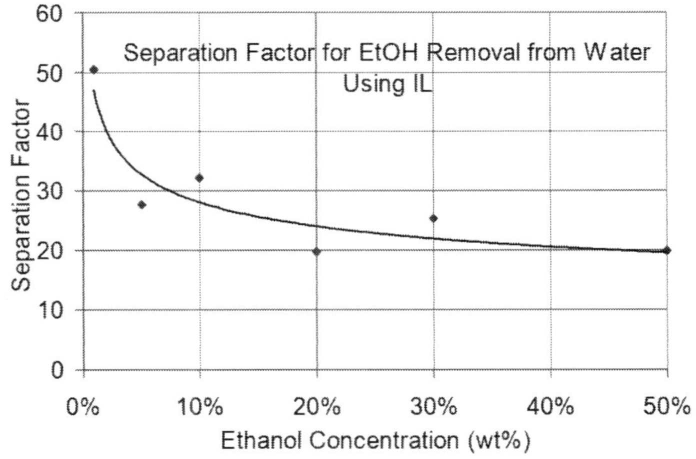

Separation factor is around 30-50 when 5% ethanol was in contact with the IL. Higher separation factor was observed when IL is used in a membrane contactor to recover 5% ethanol. More detailed results will be discussed in the presentation followed with the energy consumption calculation from process model. We will also discuss the product recovery from the IL extractant.

CO_2 Removal from CO_2/H_2 Mixtures Using Immobilized Liquid Membranes Containing Amine Blends as Carrier

Aliakbar Heydari Gorji [a], **Tahereh Kaghazchi** [a], **Ali Kargari** [a,b]

[a] Department of Chemical Engineering, Amirkabir University of Technology (Tehran Polytechnic), No. 424 Hafez Avenue, Tehran 15875-4413, Iran
[b] Department of Chemical Industries, Iranian Research Organization for Science and Technology (IROST), No. 71, Mousavi Street, Enghelab Ave., Ferdowsi Sq., P.O. Box 15815-3538 Tehran, Iran

Abstract

Selective separation of CO_2 from a wet gaseous mixture of CO_2-H_2 through facilitated transport membranes containing immobilized aqueous solutions of MEA, DEA and EDAH+ and their mixtures was experimentally investigated. The effects of CO_2 partial pressure, amine concentration and amine type on CO_2 and H_2 permeations have been studied. The CO_2 permeance through amine solution membranes decreases with increasing CO_2 partial pressure of feed and H_2 permeance was almost independent of H_2 partial pressure. Experimental results showed that the mixed or single amines with low viscosity and moderate equilibrium constant (i.e. large forward and reverse reaction rate of CO_2-amine) are proper for effective separation of carbon dioxide. Permeability of CO_2 generally increased with an increase in amine concentration howbeit it may be compromised by the salting effect and decreased species diffusivities. The results obtained indicated that permeation rate of CO_2 across a variety of amines are in order of DEA (2M)>MD (2M)>MD (1M)>MEA (2M)>MD(4M)> DEA (1M)>MEA (1M) for various concentrations of MEA and DEA mixture and are in order of $EDAH^+$ (2M)> DEA (2M)> MH (2M)>DH (2M)> MEA (2M) for various mixtures of amines.

1. Introduction

The selective removal of CO_2 form natural gas, refinery and synthesis gas streams is an important and critical process because CO_2 poisons catalysts, it is one of major greenhouse gases and highly corrosive.

Hydrogen is an environmentally clean energy source since unlike fossil fuels, it produces water as the only by product. Hydrogen is expected to be an energy source in the 21st century. Hydrogen is usually produced by reforming reaction, followed by the water gas shift reaction. CO_2 associated with hydrogen should be removed since it acts as impurity in case of H_2 production and poisons catalysts when synthesis gas used as feedstock in petrochemical processes such as ammonia plants.

It is therefore extremely important to develop more efficient methods for CO_2 separation from hydrogen rich mixtures.

Facilitated transport is a well-known process in which the reversible complexation reactions are carried out between solutes and reactive carrier and produced complex diffuses across liquid membrane thickness. Carrier molecules transport the solutes through a liquid membrane and then release them on the permeate side. This mechanism enhances the permeability and selectivity of liquid membranes. So, due to their very high permselectivity compared with conventional polymeric membranes and also low energy consumption, they have attracted the attention of many industries and researchers.

Recently, the interest in the use of mixed solvents in gas treating processes is increasing due to utilizing the advantages of each amine.

Blends of amine solvents which combine the higher reaction rate of one solvent with higher reverse reaction rate or lower viscosity of another solvent are suggested to employ as liquid films immobilized in micro porous membranes.

The purpose of this study is the experimentally investigation of CO_2 permeation characteristics through immobilized various aqueous mixtures of amines membranes. The specific objective of present study is to investigate the possibility of such membranes for CO_2 separation from hydrogen streams in terms of selectivity and permeability.

2. Experimental

2.1 Materials

The solid flat sheet membranes, hydrophilic PVDF microporous membranes (thickness 125 μm, pore size 0.1 μm, porosity 70%) supplied from Millipore Co. and hydrophobic polytetrafluoroethylene (PTFE) microporous membrane (thickness 70 μm, pore size 0.1 μm , porosity 68%) purchased from Toyo Rashi Kaisha Ltd.

MEA, DEA, EDA and HCL which purchased from Merck were used. Monoprotonated ethylenediamine ($EDAH^+$) was prepared by mixing equimolar amounts of hydrochloric acid and EDA. Aqueous solution of amines was prepared by dissolving the desired amount of amines in known volumes of distilled water.

2.2 Experimental Setup

The schematic diagram of the experimental apparatus is shown in Figure 1. A double layer polymeric flat sheet membranes system is housed in a holder designed to allow a feed stream and sweep gas to pass counter-current on opposite sides of the membrane.

The hydrophilic membrane was soaked in each aqueous amines solution for more than 12 hr to ensure complete filling of the micro pores with liquid solution. The total exposed area on each surface of the membrane was 100 cm2.

The feed gas was a mixture of CO_2 and H_2 and the sweep gas was N2. The partial pressure of CO_2 in the feed gas was changed from 0.88 kPa to 17.7 kPa since the CO_2 content in industrial synthesis gas is below 15~20 mol%.

The permeated H_2 and CO_2 gases in the sweep gas were analyzed by an online Varian Gas Chromatograph. The permeation experiments were performed at temperature 25 ± 0.5 °C.

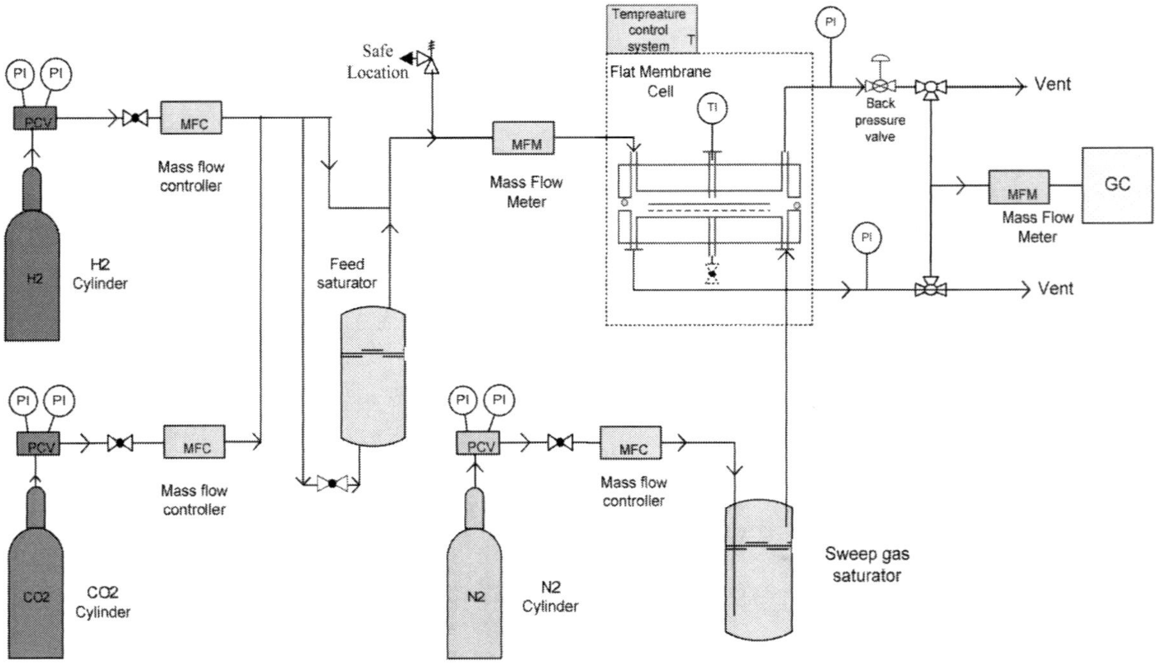

Figure 1- Schematic process flow diagram of experimental setup

3. Results and Discussion

The main objective of this research was to characterize the immobilized liquid membrane based process using mixture of amines in order to combine the benefits of each amine for effective removal of CO_2 from H_2. In this regards the permeation experiments through some conventional primary and secondary amines such as MEA, DEA, $EDAH^+$ and their mixtures in various concentrations have been performed.

The total permeation of CO_2 is composed of physically solution-diffusion plus that of chemically combined CO_2-carrier complex. Experimental results showed that the CO_2 permeance through amine solution membranes decreases with increasing CO_2 partial pressure of feed. This depression may be explained in terms of salting effect which results lesser solubility of CO_2 in amine solution. It also should be noted that the increased complex and other ionic concentrations increase liquid solution viscosity and accordingly decrease the species diffusivities.

All figures show that the permeation rate of H_2 which take place via the physically solution-diffusion mechanism is slightly decreased with H_2 partial pressure and is almost independent of H_2 partial pressure. The results obtained indicated that permeation rate of CO_2 across a variety of amines are in order of DEA (2M)> MD (2M)> MD (1M)> MEA (2M)> MD(4M)> DEA (1M) > MEA (1M) for various concentrations of MEA and DEA mixture and are in order of $EDAH^+$ (2M)> DEA (2M)> MH (2M)>DH (2M)>MEA (2M) for various mixture of amines.

A comparison of experimental results showed that the mixed or single amines with low viscosity and moderate equilibrium constant (i.e. large forward and reverse reaction rate of CO_2-amine) are adequate for improvement of selective removal of CO_2, although the H_2 permeation may be increased. Some of results have been shown in Figures 2, 3 and 4.

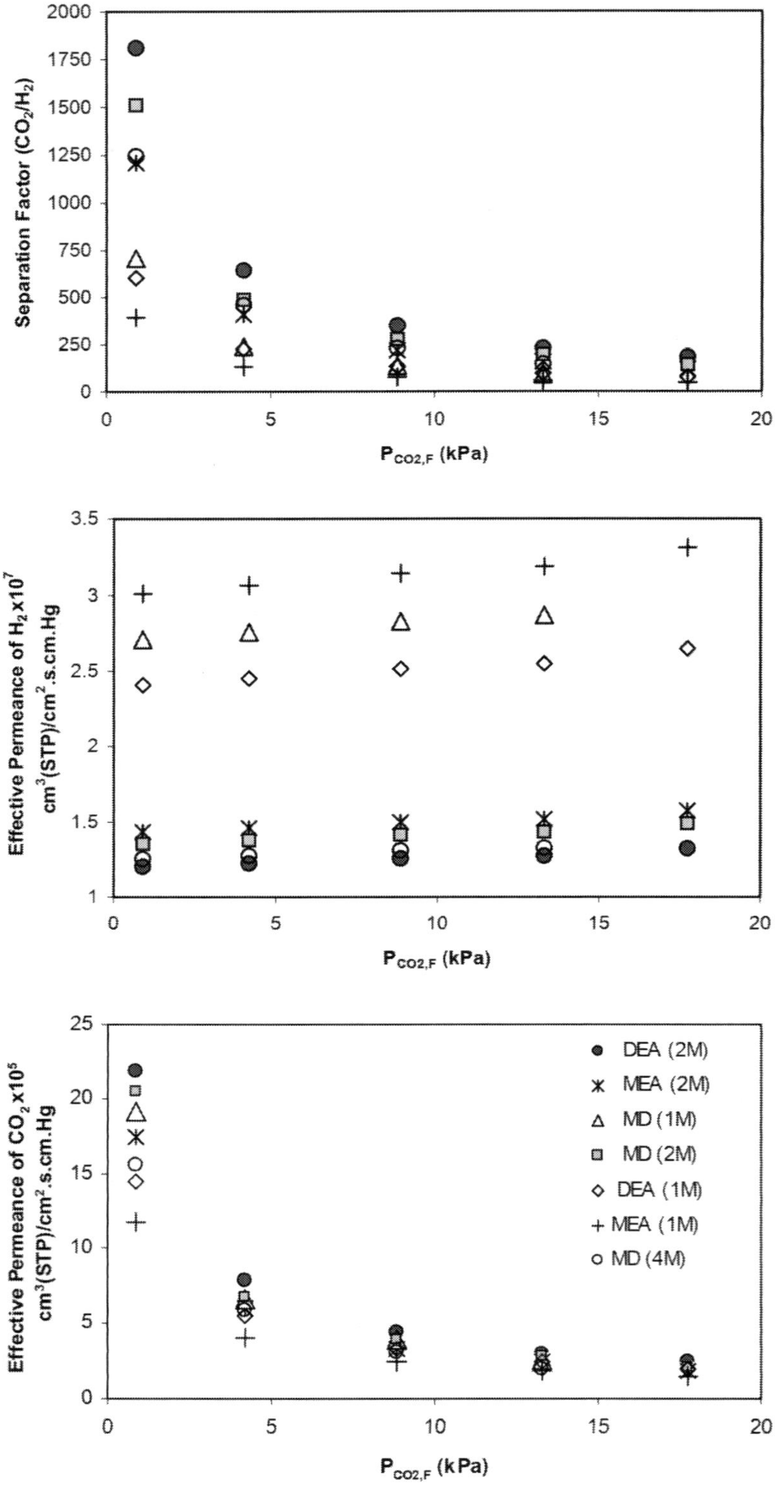

Figure 2- Effect of feed partial pressure of CO_2 on the CO_2 and H_2 permeances and separation factor for MEA(2M), DEA(2M), DEA (1M), MEA (1M), MD(1M) and MD(2M) membranes

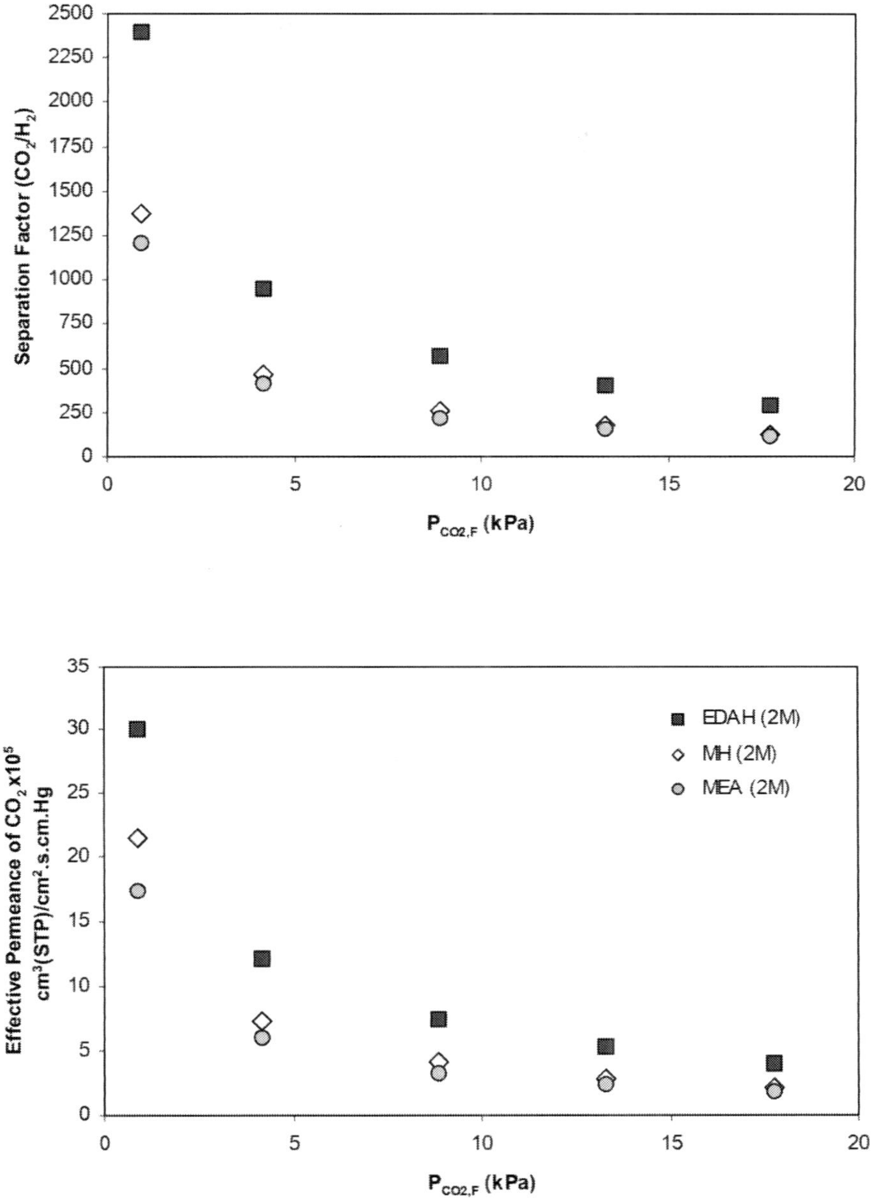

Figure 3- Effect of feed partial pressure of CO_2 on the CO_2 permeance and separation factor for MEA(2M), $EDAH^+$(2M) and MH(2M) membranes

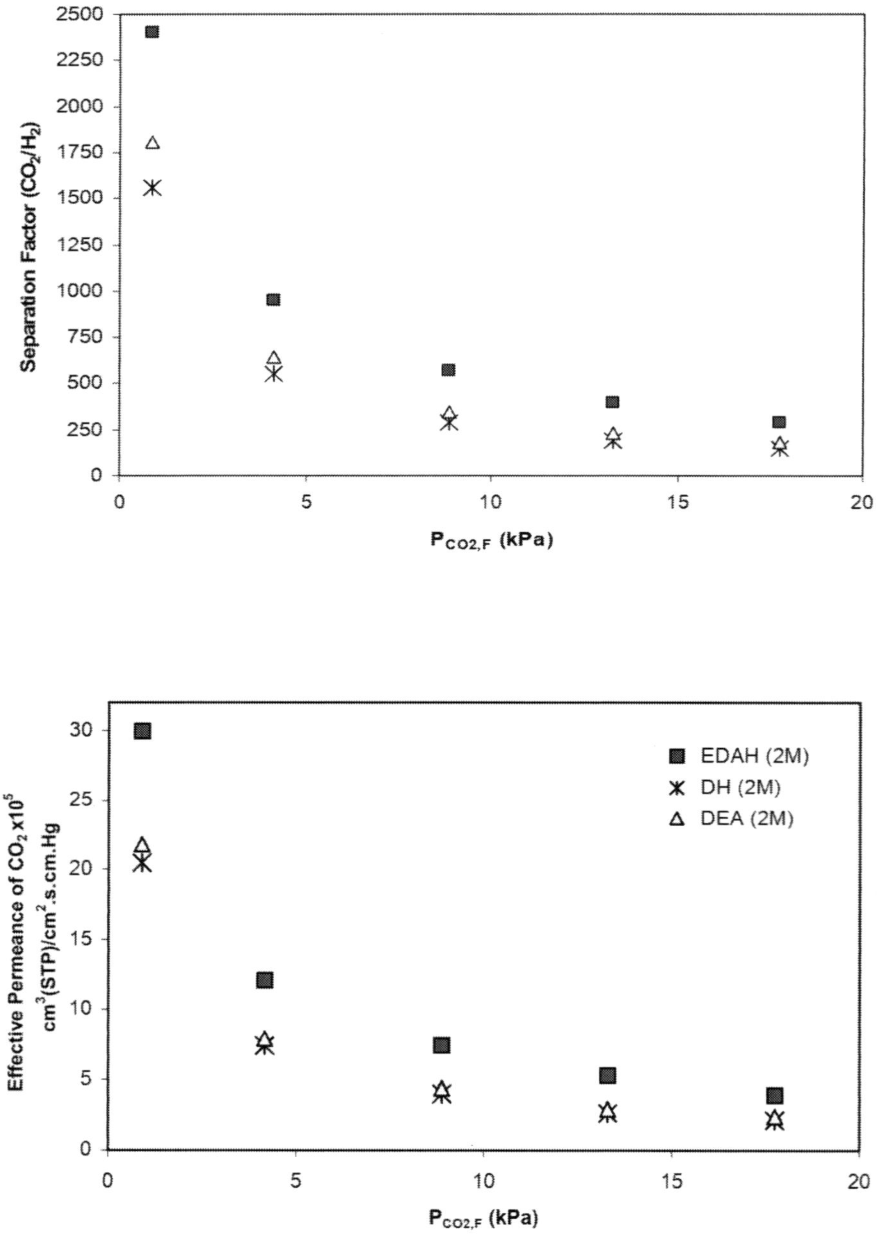

Figure 4- Effect of feed partial pressure of CO_2 on the CO_2 permeance and separation factor for DEA(2M), $EDAH^+$(2M) and DH(2M) membranes

Acknowledgements

The authors would like to acknowledge the Research and Technology Petrochemical Company for financial support of this work under the project No. 84111174.

Symbols used

M	Molar, mol/lit
P_i	Partial pressure of component i, kPa
(Q_i / L_m)	Permeance of species i, $cm^3(STP)/cm^2.s.cmHg$
DEA	Diethanolamine
DH	Mixture of Equivolume of DEA and $EDAH^+$ solutions
$EDAH^+$	Monoprotonated ethylenediamine
MD	Mixture of Equivolume of MEA and DEA solutions
MEA	Monoethanolamine
MH	Mixture of Equivolume of MEA and $EDAH^+$ solutions
Separation Factor	$\alpha_{i-j} = Q_i/Q_j$

References

[1] J.D.Way, R.D. Noble, Facilitated transport. In Membrane Handbook; W.S.W. Ho, K.K.sirkar; Eds., Chapman and Hall: New York, **1992**.

[2] L. Bao and M. C. Trachtenberg, Facilitated transport of CO_2 across a liquid membrane: Comparing enzyme, amine, and alkaline, J. Membrane Sci., **2006**, 280 , 330-334

[3] A. Ito, S. Duan, Y. Ikenori, A. Ohkawa, Permeation of wet CO_2/CH_4 mixed gas through a liquid membrane supported on surface of a hydrophobic microporous membrane, Sep. Purif. Technol., **2001**, 24, 235

[4] M. Teramoto, K. Nakai and N. Ohnishi, Facilitated transport of carbon dioxide through supported liquid membranes of aqueous amine solutions, Ind. Eng. Chem. Res., **1996**, 35(2), 538.

93c Smart Distillation Prevents Problems Before They Occur

Jimmy L. Humphrey, J. L. Humphrey & Associates, 3605 Needles Dr., Austin, TX 78746, J. Christopher Lewis, Technology R& D, ConocoPhillips, 166 PDC Bartlesville Technology Center, Highway 60 & 183, Bartlesville, OK 74004, A. Frank Seibert, Separations Research Program, University of Texas at Austin, 10100 Burnet Road, Bldg 133/CEER, Austin, TX 78758, and John P. Farone, Farone, 511 Fern Court, Cedar Park, TX 78613.

What is Smart Distillation? This question will be answered and new ideas and information will be presented on the advantages and challenges of Smart Distillation. Examples will be presented which show how unplanned shutdowns, excessive energy use, and safety problems can be prevented.

Hydrodynamic Performance of Crossflow Concentric-ring Rotating Bed

Wang, G. Q., Jiao, Y. Q., Xu, Z. C., Ji, J. B.*

Institute of Chemical Engineering and Materials Science, Zhejiang University of Technology, Hangzhou, Zhejiang 310014, Peoples Republic of China.

Abstract

A new kind of high-gravity device—crossflow concentric-ring rotating bed is developed, the rotor of which contains a series of concentric rings. Its structures and flow channel arrangements determine its super features, such as low gas pressure drop and power consumption. The gas flow passes axially through the rotor and the liquid travels radially from the inside to the outside of the rotor. During this process, the gas flow repeatedly contacts cross-currently with the liquid flow between the concentric rings. This contact mechanism is favorable to gas-liquid stripping operation widely used in chemical process industries. In a pilot crossflow concentric-ring rotating bed, the pressure drop and the power consumption were measured under different operation conditions. The experimental results show it has lower gas pressure drop and power consumption.

Keywords

Concentric-ring, Rotating bed, Hydrodynamic, power consumption

Introduction

High gravity (HIGEE) technology, as a promising alternative to enhancing heat and mass transfer, has emerged from the concepts of process intensification (PI) and is aimed towards the substitution gravitational field with a high centrifugal filed induced by rotating a doughnut-shaped rigid bed with an eye in the center[1]. In a centrifugal filed, the liquid is generally subjected to a high acceleration of several hundreds gravity, tuned by rotational speed, as it flows through the rotating bed. Due to the larger centrifugal acceleration, the liquid is easily dispersed into fine droplets with very large surface area, which then interact strongly with the gas stream to intensify the micro-mixing and mass transfer of gas-liquid two phases by one or two orders of magnitude; thus it requires smaller equipment compared with that of its conventional counterpart, which reduces the capital and operating costs.

Originally, the flow mode of gas and liquid phase in rotating bed is counterflow [2, 3]. Liquid enters at the eye of the bed and flows radially outward through the bed due to centrifugal force. Gas is introduced at the outer periphery of the rotor and forced to flow radially inward by pressure driving force. The counterflow arrangement results in two obvious disadvantages. One is larger pressure drop because of centrifugal force and variable gas flow area. The other is relative low capacity limited by the inner area of the bed. To overcome these disadvantages, rotating bed with crossflow was developed [4, 5], the rotor of which is typically packing. What the crossflow differs from counterflow is that the gas flows axially

* To whom correspondence should be addressed. Tel.: +86 571 88320053. Fax: +86 571 88320053. E-mail: jjb@zjut.edu.cn

through the bed perpendicularly to the liquid flow. However, a disadvantage of rotating packing bed is radially non-uniform density of packing after a long-term operation due to its bad dynamic balance, which has an obviously adverse effect on mass transfer performance.

In view of these considerations, a new non-packing crossflow rotating bed was developed, which is termed concentric-ring rotating bed. In this paper the structures and features of concentric-ring rotating bed was briefly described and its hydrodynamic performance as well as power consumption were experimentally investigated and presented.

Construction of rotating bed

The concentric-ring rotating bed is a novel kind of non-packing contactor based on HIGEE technology and is characterized by a rotor combining a rotating disc with a series of concentric rings fixed on it as shown in Figure 1. The rotational disc perforated with large holes as gas channels was driven by a motor through a vertical shaft. The concentric rings, the upper parts of which were perforated with small holes as liquid channels, were coaxially installed on the rotating disc with equal radial space. To prevent the gas from bypassing the rotational disc, a labyrinth seal between the rotational disc and the stationary casing is indispensable. Liquid distributor rotating with the shaft was installed at the center of the rotor to obtain an axial uniform liquid distribution. The gas is introduced by the gas inlet into the casing and flows upward axially through the rotating disc driven by pressure difference. The liquid is fed by the liquid inlet into distributor and travels radially outward through the small holes at the upper parts of concentric rings in form of fine droplets due to centrifugal force. The liquid leaves from the rotor as a shower of droplets, which is collected on the casing wall and flows through the guide tubes into liquid tanks. In the rotor the continuous gas phase flows perpendicularly to the dispersed liquid phase. Obviously the crossflow of gas and liquid flow make the liquid repeatedly contact with the gas in the space between the adjacent concentric rings. This is especially advantageous to stripping processes, which involves removal of one or more volatile components from a liquid by contacting it with a gas stream.

The concentric-ring rotating bed has some excellent features compared with other kinds of rotating beds owing to its unique structures and flow arrangements. One striking advantage is lower gas pressure drop. On one hand, the gas flow doesn't need to overcome the centrifugal force when it axially travels through the rotor. On the other hand, the pressure loss due to flow area changes in counterflow mode doesn't exist. Therefore the pressure drop of crossflow concentric-ring rotating bed mainly results from form drag exerted by rotational disc and the friction between the gas and liquid. The other advantage is lower power consumption of motor because of a very lower liquid holdup in rotor. These advantages make crossflow concentric-ring rotating bed applicable to stripping process with a large gas flowrate.

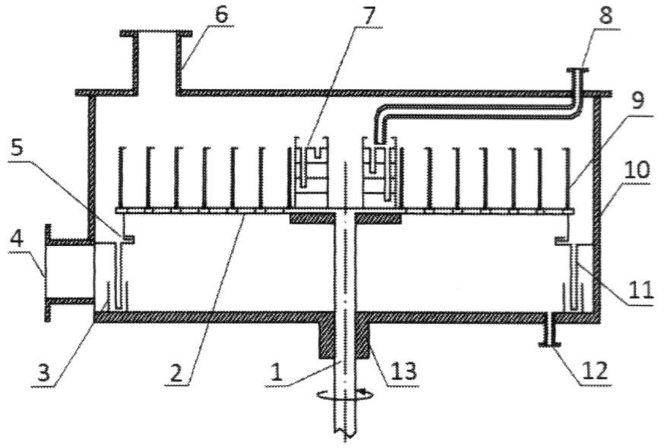

1-rotating shaft 2-rotational disc 3-liquid tank 4-gas inlet 5-labyrinth seal 6-gas outlet 7-liquid distributor 8-liquid inlet 9- concentric rings 10-casing 11-liquid guide tube 12-liquid outlet 13- shaft seal

Fig.1 Simplified sketch of crossflow concentric-ring rotating bed

Experimental Setup and Procedures

The experimental investigations were carried out in a pilot crossflow concentric-ring rotating bed using air-water system and a simplified schematic diagram of the experimental system is shown in Figure 2. The inner and outer diameter of the rotor, which is installed in a casing with 400mm in diameter, is 35 and 11 cm respectively. The axial height of the concentric rings is 6.5 cm and the radial distances between the adjacent concentric rings are 2 cm. In this device the rotor can be operated from 800 to 1400 rpm, which provides centrifugal force of 80g to 250g based on the arithmetic mean diameter.

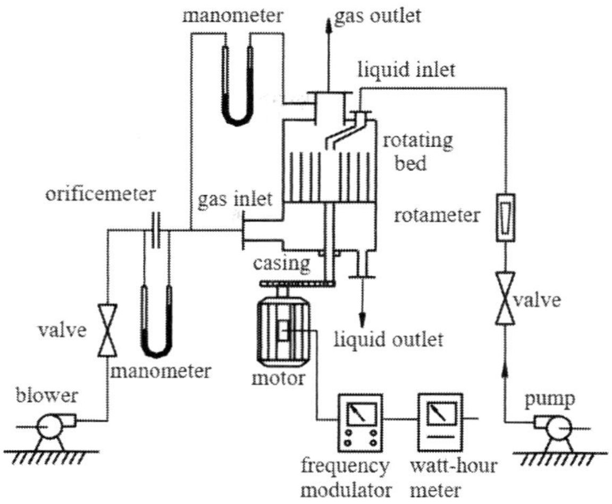

Fig. 2 Schematic diagram of experimental setup

In this experiment, the gas from blower is measured by orificemeter and introduced tangentially into the casing. The gas then flows axially upward through the rotational disc and leaves the rotor by the gas outlet. The liquid from pump enters the rotor after measured the flowrate by rotameter and then travels radially outward through the concentric rings due to the centrifugal force. The rotor is driven by a motor and the rotational speed was regulated using a frequency modulator. The pressure drop was measured by U-tube manometer and the power consumption of motor was determined by three-phase Watt-Hour meter and stopwatch. The procedures are measuring the pressure drop and power consumption under different gas flowrates at constant liquid flowrate and rotational speed. Then after the liquid flowrate or rotational speed is changed the procedures were repeated to study the pressure drop and power consumption dependence on gas and liquid flowrate as well as the rotational speed.

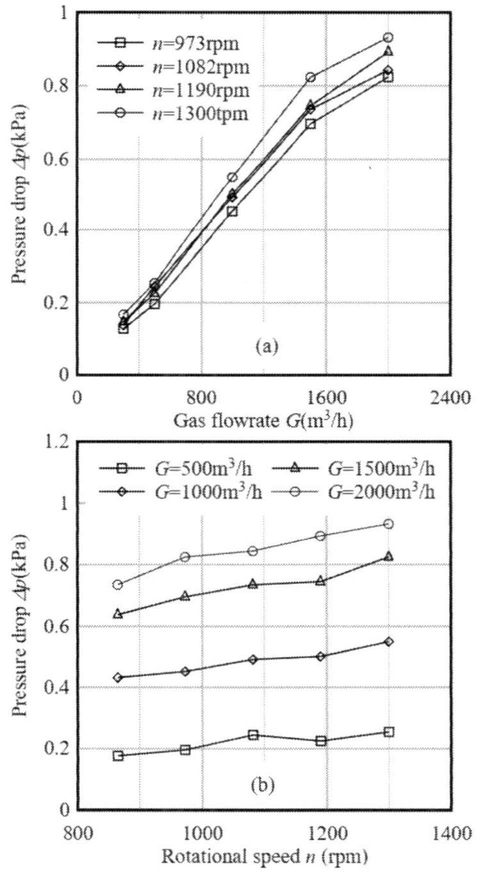

Fig.3 Variation of the dry pressure drop (a) with gas flowrate (b) with rotational speed

Experimental Results and Discussions

Gas Pressure Drop

The effect of the gas flowrate and the rotational speed on the dry pressure drop across the rotor in the absence of liquid is shown in Figure 3a and 3b. It can be seen from Fig. 3a that the dry pressure drop increases with increasing gas flowrate because of the fact that the dry pressure drop mainly results from the friction force and form drag of gas across the rotor which increases with the gas velocity increase. Fig.3b shows that the dry pressure drop slightly increases with the increase of the rotational speed. On one hand, the centrifugal force has almost no effect on gas flow because the gas flows perpendicularly to the centrifugal filed. On the other hand, the friction between the gas and concentric rings increases a little with rotational speed increase, which results in a slight increase of the dry pressure drop.

The dependence of wet pressure drop on gas and liquid flowrate, liquid flowrate as well as the rotational speed was shown in Figure 4a, 4b and 4c. It can be seen from Fig.4a that with gas flowrate increase the wet pressure drop obviously increases at a constant rotational speed and liquid flowrate. This variation is similar to that of the dry pressure drop. Fig.4b shows that the wet pressure drop is nearly independent of liquid flowrate. Generally in centrifugal filed the liquid is spread into fine droplets leading to very small liquid holdup. Therefore increase in liquid holdup and wet pressure drop is very small in spite of an obvious liquid flowrate increase. This feature is a striking advantage of rotating bed compared with conventional tower. Fig.4c indicates that the rotational speed has a moderate effect on the wet pressure drop at different gas and liquid flowrate. It can be seen from the comparison between Fig.3b and Fig. 4c that the increase of wet pressure drop with rotational speed is larger than that of dry pressure drop. This difference can be explained by the interaction between liquid phase and concentric rings that is enhanced by increasing rotational speed. In addition, it is worth while to note that the wet pressure drop of concentric-ring rotating bed is usually lower than 1.0kPa, which could meet the pressure drop requirement of gas absorption and stripping process.

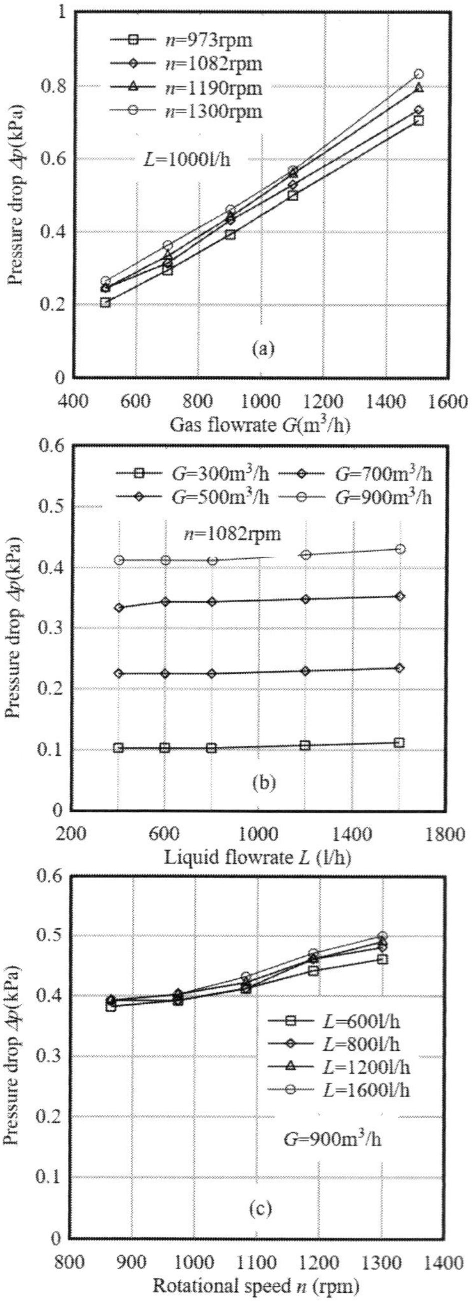

Fig.4 Variation of the wet pressure drop (a) with gas flowrate (b) with liquid flowrate (c) with rotational speed

Power Consumption

In a rotating bed, power is required to accelerate the liquid and to overcome bearing friction and frictional losses as the liquid passes through the bed. Moreover, whether a rotating bed is energy-saving depends on its power consumption, which also determines the choice of motor. The power consumption of rotating bed depends on gas flowrate, liquid flowrate, rotational speed and structures of rotor. In this study, the effects of gas and liquid flowrate as well as rotational speed on power consumption were experimentally investigated and presented below.

The variation of the power consumption with gas and liquid flowrate, liquid flowrate as well as the rotational speed was given in Figure 5a, 5b and 5c. It is indicated by Fig. 5a that power consumption increases with increasing gas flowrate at constant liquid flowrate and rotational speed. Gas flowrate increase usually leads to more liquid holding in bed and more power from motor to accelerate it. At lower rotational speed, gas flowrate has little effect on power consumption whereas this effect becomes more obvious at higher rotational speed. Fig. 5b suggests that increasing liquid flowrate can increase the power consumption at certain gas flowrate and rotational speed. The increase of liquid flowrate can increase the energy losses due to impingement of liquid droplets on concentric rings and acceleration of these droplets again. Fig. 5c reveals that the higher rotational speed can bring an obvious increase of power consumption. This can be attributed to the energy losses resulting from liquid impingement on concentric rings with higher velocity due to higher rotational speed and from bearing friction.

Although the power consumption is relatively small (about 1.0kW) based on our experimental results, it still seems that compared with conventional columns, rotating bed is at a disadvantage due to this power consumption. But the reduction in size and consequent

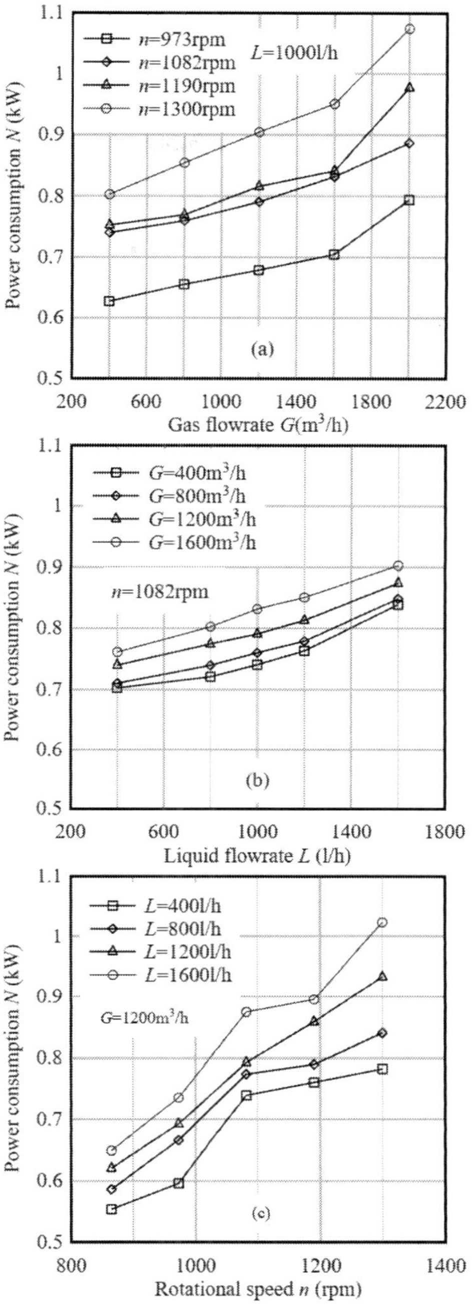

Fig.5 Variation of the power consumption (a) with gas flowrate (b) with liquid flowrate (c) with rotational speed

lower capital costs might be sufficient to offset the additional costs. However, this issue needs an overall evaluation.

Concluding Remarks

As new kind of high-gravity equipment, cross-flow concentric-ring rotating bed is developed with a rotor containing a series of concentric rings. This device possesses striking advantages such as low pressure drop and power consumption. In the rotor, the gas flows cross-currently and repeatedly contacts with the liquid flow between the concentric rings, which is especially beneficial to gas-liquid stripping operation. In a pilot rotating bed system, the pressure drop and the power consumption were studied under different operation conditions. The experimental results show that with increasing gas flowrate, the pressure drop increase, and the liquid flowrate and the rotational speed has a slight and moderate effect on pressure drop respectively. And the reduction of power consumption mainly depends on the decrease of gas and liquid flowrate as well as the rotational speed. But the effect of the liquid flowrate is less obvious than that of the gas flowrate and rotational speed. Cross-flow concentric-ring rotating bed has pressure drop about 1.0kPa and power consumption about 1.0kW. This relatively low energy consumption shows it is a promising choice in chemical industries.

Acknowledgements

This Project was supported by Zhejiang Provincial Natural Science Foundation of China through grant Y406239, which is gratefully acknowledged.

References

[1] Ramshaw C. HIGEE distillation-an example of process intensification. *The Chem. Engr.*, 1983, 389, 13-14.

[2] Kumar, M. P., Rao, D. P. Studies on a High-Gravity Gas-Liquid Contactor. *Ind. Eng. Chem. Res.*, 1990, 29(5), 917-920.

[3] Liu, H. S., Lin, C. C., Wu, S. C., Hsu, H. W. Characteristics of a Rotating Packed Bed. *Ind. Eng. Chem. Res.*, 1996, 35(10), 3590-3596.

[4] Guo, Fen; Zheng, Chong; Guo, Kai; Feng, Yuanding; Gardner, Nelson C. Hydrodynamics and mass transfer in cross-flow rotating packed bed. *Chem. Eng. Sci.*, 1997, 52(21/22), 3853-3859.

[5] Lin, Chia-Chang; Wei, Tzu-Ying; Hsu, Shu-Kang; Liu, Wen-Tzong. Performance of a pilot-scale cross-flow rotating packed bed in removing VOCs from waste gas streams. *Sep. Purif. Technol.*, 2006, 52(2), 274-279.

35c Stability and Selectivity of Supported Liquid Membranes with Ionic Liquid for Separation of Organic Liquids by Vapor Permeation

Baoguo Wang, Zhaorong Zeng, Jia Lin, Feng Wu, Li Ding, and Yong Peng. Department of Chemical Engineering, Tsinghua University, Beijing, 100084, China

Room temperature ionic liquids (RTILs) have attracted a great deal of attention as green solvents, and are used in many fields of the chemical industry, because of their thermal stability and negligible vapor pressure, which makes them environmentally friendly. Supported liquid membranes (SLMs) are often prepared by immobilizing organic solvent inside a porous structure of polymeric or ceramic membranes, and become a promising alternative to conventional solvent extraction, since they combine the operations of extraction and stripping. However, SLMs have not obtained commercial application in the past 20 years, mainly due to their instability and poor long-term performance. In order to prevent the loss of membrane liquids, we propose a novel approach in which vapor permeation with an ionic liquid filling-type SLM replaces solvent extraction. Since the basic characteristics of RTILs are high surface tension and a lack of detectable vapor pressure, this method provides several possible advantages: no adjacent liquid phase contacts the ionic liquid directly, thus preventing the loss of membrane liquid by the dissolution/dispersion effect, as well as by evaporation. Moreover, the molecular structure of ionic liquid used for SLMs can be designed corresponding with the required physicochemical properties of the desired permeation component in the organic mixtures. The separation of toluene/cyclohexane mixtures was conducted for as long as 550 hours, and the separation factor was kept within the range of 15 to 25, showing good durability in vapor permeation. The use of SLMs with ionic liquid was shown to be a promising method to separate organic/organic or organic/water mixtures.

229c Novel Test Results Demonstrate Exciting Advantages for Raschig Super-Pak

Michael Schultes, Raschig Ring Department, RASCHIG GMBH, Mundenheimerstr. 100, Ludwigshafen, Germany and Simon Chambers, Raschig-Ringe, RASCHIG GMBH, Mundenheimerstr. 100, Ludwigshafen, Germany.
Raschig Super-Pak is a new generation of structured packing with a novel surface geometry. Raschig Super-Pak enables, to an extent never known before, high separation efficiency and high loading capacity while keeping the pressure drop extremely small. This structured packing is fundamentally different to the standard and high capacity corrugated sheet metal structured packings that have dominated the market over the past 30 years. Raschig Super-Pak has an open structure such that vapor-liquid traffic can flow freely within a packing element and no sharp directional changes are existing at the layer interface. The rows of sinusoidal waves within vertical packing sheets are surface enhanced to encourage greater turbulent radial spread of thin liquid film flows on the front and back of the waves on each sheet within an element. New test results will be presented and the performance will be compared to existing geometries on the market.

163c Simplified Modelling of Crude Oil Distillation Columns Meeting Refining Product Specifications

Lu Chen, Megan Jobson, and Robin Smith. Centre for Process Integration, School of Chemical Engineering and Analytical Science, The University of Manchester, Manchester, United Kingdom

The design and optimisation of crude oil distillation columns is of great importance in the refining industry. The process consumes considerable amount of energy and the refining product properties are mainly determined by the performance of the separation.

Simplified models offer a number of benefits for the design and analysis of crude oil distillation columns, compared with rigorous simulation, in terms of their flexibility and robustness. Simplified models have been developed to facilitate grassroots and retrofit design and analysis of crude oil distillation columns (Suphanit, 1999; Gadalla, 2003; Rastogi, 2006). These models require the specification of key components and their recoveries to define the separation. These models perform well, compared to rigorous simulations. However, due to the large number of components in crude oil, even with application of pseudo-components to dramatically reduce the number of components, making an appropriate choice of key components and their associated recoveries becomes a problem when applying the simplified models for the design and analysis of crude oil distillation columns. Previous approaches have required preliminary rigorous simulation results, and have used trial and error to determine the most appropriate key components and recoveries to give a good match between rigorous and simplified models.

Conventionally, refining products are specified in terms of true boiling properties, bulk properties, such as Specific Gravities, Sulphur Content, Refractive Index, Watson K, Research Octane Number, Viscosity, Reid Vapour Pressure, Flash Point, Cetane Index, etc.. and flow rates. In this paper, a novel methodology is proposed to automatically identify appropriate key components and recoveries given conventional product specifications. This new approach enables the use of simplified models to solve industrial relevant design problems.

Case studies illustrate the approach, and compare the results of simplified models with those of more rigorous models. These simplified models can then be coupled with heat exchanger network models to allow the simultaneous optimisation of the column and heat recovery system.

Reference

M. Gadalla, M. Jobson, and R. Smith, Shortcut models for retrofit design of distillation columns, Trans IChemE, September, 81(A), 971-986 (2003)

Rastogi, V., Heat Integrated Crude Oil Distillation System Design, PhD Thesis, The University of Manchester, Manchester, UK (2006)

Suphanit, B., Design of Complex Distillation System, PhD Thesis, UMIST, Manchester, UK (1999)

229b Distillation Tower Operation Experience

Daniel R. Summers, Sulzer, 4019 South Jackson Avenue, Tulsa, OK 74107

A trayed distillation tower experince will be presented. The experience will be enlightening and useful to the audience. Operating data will be presented and results will be given.

200a Implementation and Validation of a Model-Centric Support System within a Pilot Plant Scale Packed Distillation Column

Daira Aragon[1], Pablo A. Rolandi[2], and Jose A. Romagnoli[1]. (1) Chemical Engineering, Louisiana State Univeristy, Cain Department of Chemical Engineering, South Stadium Road, Baton Rouge, LA 70803, (2) gPROMS Kernel and Solvers Team, Process Systems Enterprise Limited, Bridge Studios, 107a Hammersmith Bridge Road, London, United Kingdom

Separation of mixtures is an important stage in the production process of a considerable number of products. In most cases, the quality and purity of the final products depend on the degree of separation achieved. Therefore, special attention has been drawn to distillation of vapor/liquid mixtures as it is a typical operation in petrochemical and chemical industries. During the normal operation of a distillation unit at plant scale, temperature and flow measurements are collected through the distributed control system (DCS). This information may serve for many different purposes such as process control, parameter estimation, data reconciliation and optimization, among others. However, the majority of studies in the area show the use of such information for control and optimization purposes only.

The conventional equilibrium stage model has been commonly utilized. It assumes that the phases leaving each stage are at complete equilibrium, simplifying the model of tray distillation in comparison to the description of the process in terms of continuous mass and energy balances (Perry, R.H.). This approach has been also applied to packed distillation columns assuming they are composed by several sections, each of them being treated similarly to a stage in a tray column (Steiner et al., 1978; Krishnamurthy, R. and Taylor, R., 1985; Yang, L. and Chuang, K.T., 2000). Fortunately, new methodologies for the solution differential and algebraic equations (DAEs) and the increasing computational capability of computers have originated the development of models based on first principles for packed distillation units assuming that the mass transfer in continuous across the column are being developed (Karacan, S. et al., 1998; Fernández-Seara et al., 2002; Alpbaz et al., 2003). It is possible now to think in the application of other advanced model-based methodologies such as dynamic data reconciliation and parameter estimation to packed distillation processes using more fundamental modeling.

Concurrently, the consolidation of the CAPE (Computer Aided Process Engineering) community in the 1990's, the subsequent development of the CAPE-OPEN (CO) project and the advances in state-of-the-art modeling environments paved the way for a paradigm shift towards technologies based on models. The conceptual definition of a single and consistent model-centric framework for integrated decision support of process systems (IDSoPS) was proposed in previous work (Rolandi and Romagnoli, 2005; Aragon et al. 2006). It provides an innovative approach for simplifying the problem formulation of model-based activities (i.e. simulation, parameter estimation, data reconciliation and optimization) by incorporating the Problem Definition Environment (PDE), which takes advantage of the capabilities of state of the art modeling and solution environments (MSEs).

In this work, the mechanisms and methods employed in the IDSoPS and its corresponding PDE are presented and discussed through a case study of a pilot plant packed distillation unit. The backmixing model is assumed to develop a rigorous dynamic model based on first principles for the continuous operation of the packed distillation column with a ternary feed composed. In addition to mass and energy balances, mass and heat transfer along the tower are included in the model. A state-of-the-art modeling environment, capable of solving distributed and highly nonlinear equations on-line in non-stationary state will be used to implement and test a number of advanced operational activities such as: joint data reconciliation (dynamic) /parameter estimation, troubleshooting of past operating conditions and dynamic optimization studies (transition planning).

A New Model for Predicting Gas / Liquid Effective Surface Area in Packing

J. Christopher Lewis, A. Frank Seibert, James Fair

Accurate measurement of the interfacial area of random and structured packings will allow better models to be created for use in designing distillation and absorption columns. Researchers at the University of Texas at Austin have developed a procedure that allows the effective interfacial area of random and structured packings to be calculated directly from experimental results. These data are presented and compared to existing interfacial area models. Furthermore, a new model is presented for accurate estimation of the effective interfacial area.

Eddy Diffusivity Spiegler-Kedem (EDSK) Model for Reverse Osmosis Membrane

Senthilmurugan Subbiah[*a] and Shard K Gupta[b]*
[a] *Corporate Research Centre, ABB Global Services Ltd, Bangalore, KA, India.*
Tel: +91 80 2294 8298, Email: senthilmurugan.s@in.abb.com
[b] *Chemical Engineering Department, Indian Institute of Technology Delhi, ND, India.*
Fax: +91-11-26581120; Tel: +91-11-26591023; Email: sgupta@chemical.iitd.ac.in

Introduction

Reverse osmosis (RO) and nanofiltration (NF) are important techniques in water and waste water industries. The RO and NF membranes were characterized by many people [1-11] from 1970 itself by solving mass transport equation of appropriate membrane transport model and film theory for mass transport in film. According to Spiegler-Kedem model, the observed rejection goes from zero to maximum value and to zero again as the feed pressure increases in both RO and NF (Opong et. al. [3], Murthy et al. [1,2]). In fact, Murthy et al. [1,2] has mathematically shown that this maximum in the observed rejection occurs at some $J_v |_{\text{at } R_{Max}}$ given by the following equation.

$$J_v\Big|_{R_{Max}} = kp_m \left[\frac{\ln\left(1 + \frac{(1-\sigma)k}{p_m}\right)}{(1-\sigma)k} \right] \quad \alpha \quad \Delta p \qquad (1)$$

Now, depending on the side of the curve on which the data is observed, the observed rejection may increase or decrease as the feed pressure is increased. This behaviour in observed rejection has not been reported in the literature for NaCl separation for reverse osmosis membranes. In the recent investigations [4] this behaviour was observed for Nanofiltration. This behaviour may be due to the influence of J_v in mass transfer coefficient [4-9]. A general equation for the mass transfer coefficient is given by

$$Sh = a(\text{Re})^b (Sc)^{1/3} \qquad (2)$$

The value of a and b depends on module configurations and flow pattern. The values of a and b for different geometry and flow pattern was summarized by Gekas et al. [9]. On the other hand, the correlation for mass transfer coefficient for high flux mass transfer in both RO and NF was reported by many people [4-9] as a function of permeate flux on the basis of the eddy diffusivity model at high flux rate (J_v). The general form of this correlation is given below.

$$Sh = a(\text{Re})^b (\text{Re}_p)^c (Sc)^{1/3} \qquad (3)$$

where, Permeation Reynolds number = $\text{Re}_p = \frac{\rho_p D J_v}{\mu_p}$,

Circulation Reynolds number = $\text{Re} = \frac{\rho_F D v}{\mu_F}$,

The value of a and b were estimated from the experimental data. The estimated literature values of a and b for both RO and NF membranes are shown in Table 1. Thomas [8] concluded that if the permeate flux exceeds more than 0.3 cm/min and Schmidt number should be around 600 then the eddy diffusivity model is most suitable for describing liquid film mass transport in hyper-filtration. In order to obtain the range of Schmidt number, the experiments were carried out for $MgCl_2$ at 10 °C and HCl at 27 °C resulting with a Schmidt number of 1850 and 280. Also recommended that for higher permeate flux rate, the film theory was not applicable. Wang [8,9] reported the dimensionless equation for calculating the concentration polarization under the condition of steady state and turbulent flows as given in Eq.(4). Afonso et al, [5] derived the equation for estimation of membrane surface concentration on the basis of eddy diffusion model. The final equation for the estimation of mass transfer coefficient is given below.

Table 1: Literature values of constant used in mass transfer coefficient correlation equation (4)

Reference	a	b	c	d	Remarks
Afonsob et al [4,5]	0.162	0.29	0.93	0.667	The values determined for NaCl-water system
David G. Thomas [7]	*	0.188	0.5	0.25	* The value of a is reported as function of ratio between feed velocity and permeate flux
Wang Xuesong [8,9]	**	0.74	0.81	-0.85	** The value of a is reported as a function dimension of test cell

$$\frac{C_m - C_p}{C_b - C_p} = \exp\left(\frac{a\left(\mathrm{Re}_p\right)^c (Sc)^d}{(\mathrm{Re})^b}\right) \qquad (4)$$

They used extended Nernst Planck equation for describing the membrane transport and eddy diffusivity model for estimating the membrane surface concentration. The experiments were carried out for the separation of NaCl from the aqueous solution in flat sheet nanofiltration membrane. The value of a, b, c and d are estimated by curve fitting of experimental values and model prediction. The estimated value of a=0.162, b=0.26, c=0.93, d=0.667. The rejection of sodium chloride was found to have maximum with respect to feed pressure for low feed flow rate and concentration. The model was found to be in good agreement with experiments. Also Afonso et al, [4] extended the experiment for single solute separation of $MgSO_4$, $MgCl_2$, Na_2SO_4 from the aqueous solution by using commercial flat sheet nanofiltration membrane. The rejection of all the three components was found to have maximum with feed pressure for low feed flow rate. The model described in [5] was used to theoretical prediction of all three components. The relative errors between the experimental and calculated permeation fluxes and the salt rejection were lower than 22 and 12 % respectively.

In another study, Geraldes et al. [6] developed a numerical model based on the finite volume formulation to predict laminar flow hydrodynamics and mass transfer of aqueous solution in plate and frame system. They reported that the film thickness is a function of Schmidt number, and circulation and permeation Reynolds numbers. Ultimately the final equation for mass transfer coefficient will be in the form of equation (3). The experiments for separation of NaCl, Na_2SO_4 and glucose from aqueous solution were carried to validate their model. The rejection of all three components was found to have maximum with respect to feed pressure. The experimental data of observed rejection and permeate fluxes were in excellent agreement with model predictions.

From the literature, mass transport had been explained through a combination of eddy diffusivity model for film and solution diffusion model and extended Nernst Planck equation for membrane. The parameters in equation (3) had been estimated by using this combination. But Murthy et al. [1,2] developed the CFSK model and compared it with the existing Combined Film Solution Diffusion model (CFSD) and concluded that Spiegler-Kedem model is better to explain mass transport in membrane in comparison with Solution Diffusion model. In the present work, the eddy diffusivity model for mass transport in the film was combined with Spiegler-Kedem model for mass transport in the membrane. The experiments were carried out in a membrane test cell for the separation of NaCl and $CaCl_2$ by using a commercial flat sheet membrane (Dow USA). The experimental results were analysed and the parameters of the Spiegler-Kedem model and the constants of the eddy diffusivity model were estimated by the error minimisation technique- Simplex Search.

Theory

According to Spiegler-Kedem model the equation for the permeate flux and rejection for the reverse osmosis or nanofiltration membrane is given below

Permeate flux:

$$J_v = \frac{A}{\rho}\left[(P_b - P_p) - \sigma\left[\frac{vRT}{M_w}\right](C_m - C_p)\right] \qquad (5)$$

Rejection:

$$R_T = 1 - \frac{C_p}{C_m} = \frac{(1-F)\sigma}{1-\sigma F} \qquad (6)$$

Equation (6) is combined with concentration polarization model and finally, we obtain the following equation for the concentration of the solute in the permeate stream.

$$C_p = \frac{C_b}{1 + \frac{(1-F)\sigma}{(1-\sigma)\phi}} \qquad (3)$$

Where $$F = \exp\left\{-\frac{J_v(1-\sigma)}{P_m}\right\} \qquad (4)$$

and $$\phi = \exp\left[\frac{J_v}{k}\right] = \frac{C_m - C_p}{C_b - C_p} \qquad (5)$$

The value of mass transfer, k can be estimated from eddy diffusive model as give in equation (3)

Experimental procedure and details

The inorganic salts used in the present work (NaCl and $NaSO_4$) were imported from MERCK is circulated in the experimental setup which is made of stainless steel. The schematic diagram is given in reference [1]. The setup consists of a test cell (where the feed solution enters and leaves the in the upper side of the membrane, the membrane bought from Dow Co. is placed over two sieves (100 μ and 300 μ) which are just above the porous plate), pump, feed tank, refrigeration unit, rotameter, pressure regulator, valves and a temperature controller. The feed is pressurised from feed tank by using the high pressure reciprocating pump into the test cell and then the reject stream send back to feed tank. The feed temperature was maintained by refrigeration unit fixed in between the test cell and feed tank. The membrane was treated with double distilled water for 8 hours at a feed pressure of 20 bar. After the treatment, the pure water flux was estimated for different feed pressure. In case of single solute separation, the experiments are carried out for constant feed flow rate and different feed pressure to measure the permeate flux and concentration. Then the parameters like feed flow rate and concentration are varied and the corresponding permeate flux and the permeate concentration were determined for each case. The variations in operating conditions are shown in the Table 2. The concentration of both NaCl and $CaCl_2$ is estimated by using Ion Chromatography (Metrohm 792 Basic IC)

Table.2 Variation of operating conditions

Operating conditions	Feed Concentration	Feed flow rate	Feed pressure
Range	1000, 2000 and 3000ppm	600, 1200 and 1800 ml/sec	5, 10, 15, 20 and 25 bar

Experimental results and Discussion

The rejection of NaCl and $CaCl_2$ increases and then decreases with increase in feed pressure at constant feed flow rate. The same behaviour has been obtained for different feed flow rate and feed concentration as shown in Figure 1. In general, the rejection of solute in reverse osmosis membrane increases to a maximum value and then decreases to zero as discussed in Murthy et al [1, 2]. The same behaviour had been obtained by Afonso et al [4, 5 and 6] for Nanofiltration membrane. The rejection increases with the feed flow rate at constant feed pressure, since mass transfer co-efficient increases with increase in feed flow rate and hence the concentration polarisation decreases. As shown in Figure 2 and 3, the rejection of $CaCl_2$ decreases with increase in feed concentration, on the other hand the NaCl rejection decreases first and then increases with increase in feed concentration. This type of behaviour was also reported by Soltanieh et al. [11] in reverse osmosis flat sheet membrane for NaCl system. They found that the solute permeability (diffusivity/partition coefficient) of membrane may change with respect to the salt concentration. The rejection of $CaCl_2$ is higher than the NaCl rejection for same feed conditions due to the trivalency of the $CaCl_2$. The permeate flux increased with increase in feed pressure.

Parameter estimation and discussion

The parameters of the Spiegler-Kedem model and the eddy diffusivity model is estimated for both experimental data reported in this paper and pervious literature data [2,4]. The down hill simplex method was used to estimate these model parameters by minimizing the error between the experimental and theoretical values permeate flux and concentration. The pure water hydrodynamic permeability of membrane used in the present experimental work is estimated by using pure water flux data and the estimated value of hydrodynamic permeability (A) is $7.1715*10^{-12}$ m^3/m^2.sec.Pa. The remaining parameters of Spiegler-Kedem model and mass transfer coefficient are estimated by using separation data of $CaCl_2$–water system. The estimated model parameter values are shown in Table 3. The estimated mass transfer coefficient correlation is given below

$$k = 4.24463 * 10^{-5} * Q_F^{0.3216} J_v^{0.5366} \tag{7}$$

Table 3: The estimated EDSK model parameter of $CaCl_2$–water system at constant feed flow rate and concentration data sets of P_F=5.1from 21.3 bar.

C_F	Q_F	P_m x 10^7	Sigma
Ppm	ml/min	m/sec	(-)
1161	1200	0.1442	0.989170
1161	1500	0.1490	0.990664
1161	1800	0.1552	0.991882
1926	1200	0.1255	0.983127
1926	1500	0.1370	0.986375
1926	1800	0.1455	0.988334

By equating the equation (3) and (8) the value of b and c is equal to 0.3216 and 0.5366. The equation (8) is also verified for data set with different feed concentration by estimating the model parameters as shown in Table 3 &4. The solute permeability of calcium chloride decreases as feed concentration increases and reflection coefficient is found to be constant with respect to feed concentration. The theoretical prediction is found to be in good agreement with experimental values as shown in Figure 1.

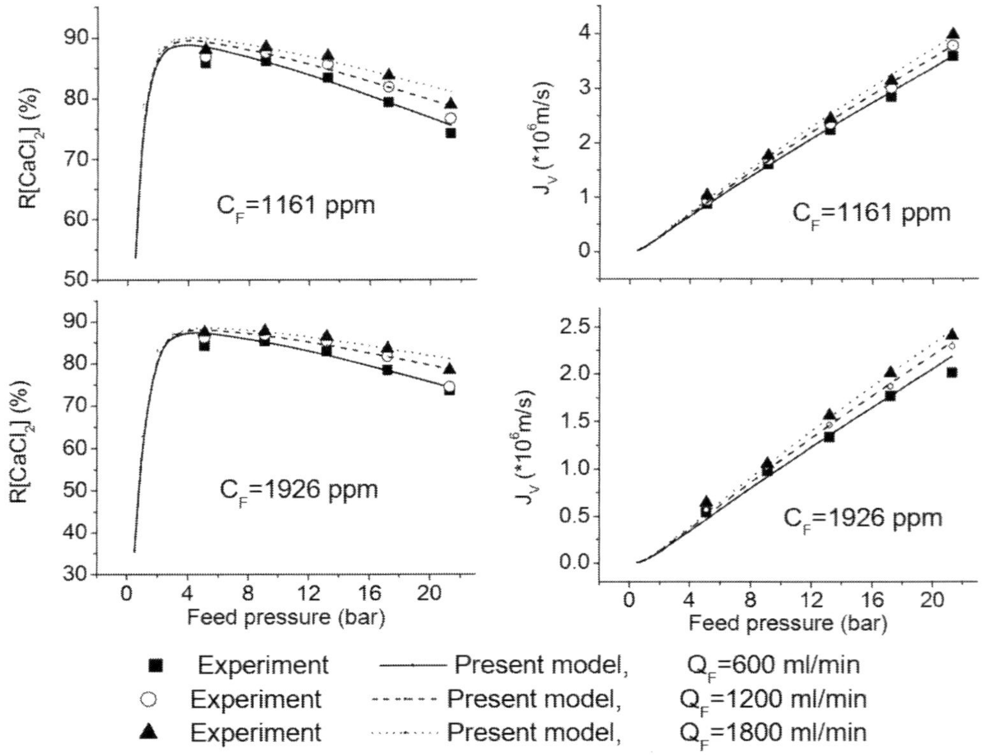

Figure 1: The permeate characteristics of $CaCl_2$ –Water system

For NaCl- water system, the estimated parameters of Spiegler-Kedem model are given in Table 4. Similar to $CaCl_2$ – water system, the estimated final correlation for NaCl- Water system is given below.

$$k = 8.306876 * 10^{-5} * Q_F^{0.3216} J_v^{0.5366} \qquad (8)$$

Table 4: The estimated EDSK model parameter of NaCl–water system at constant feed flow rate and concentration data sets of P_F=5.1from 21.3 bar.

C_F	Q_F	P_m x 10^7	Sigma
ppm	ml/min	m/sec	(-)
1335.9	600	0.3639	0.975039
1335.9	1200	0.3692	0.975715
1335.9	1800	0.3562	0.978441
1906.9	600	1.0713	0.972738
1906.9	1200	1.0918	0.978402
1906.9	1800	0.9653	0.981053
3033.7	600	0.6354	0.944595
3033.7	1200	0.6337	0.95057
3033.7	1800	0.5826	0.954985

The solute permeability of sodium chloride increases and then decreases as feed concentration increases as shown in Table 4. The reflection coefficient is found to be almost constant for both 1000 and 2000 ppm but it decreases for 3000 ppm of feed concentration. The maximum error between the experimental and model prediction values of both permeate flux and concentrations were found to be 13 and 22 % respectively

To compare the prediction of the Eddy Diffusivity and Film Theory model, the constants used in the film theory model (a and b) were also estimated in equation (2). The corresponding estimated values of Spiegler-Kedem model for $CaCl_2$-Water system are shown in Table 6. The mass transfer coefficient correlation is given below

$$k = 0.0169 * 10^{-6} * Q_F^{0.5442} \quad (9)$$

Table 6: The estimated CFSK model parameter of $CaCl_2$–water system at constant feed flow rate and concentration data sets of P_F=5.1 to 21.3 bar

C_F	Q_F	P_m x 10^7	Sigma
ppm	ml/min	m/sec	(-)
1161	1200	0.3080	0.999999
1161	1500	0.2972	0.999999
1161	1800	0.3072	0.999999
1926	1200	0.2455	0.995287
1926	1500	0.2739	0.999999
1926	1800	0.2762	0.999999

The maximum error between the experimental and theoretical values of permeate flux and concentration is found to be 45 and 56 % respectively. Also it is clearly seen from the Figure 2, that the prediction of permeate flux by CFSK model is totally different from the trend followed by the experiments. The result shows that, the combination of Spiegler-Kedem model and eddy diffusivity (EDSK) model predicts much better than CFSK model.

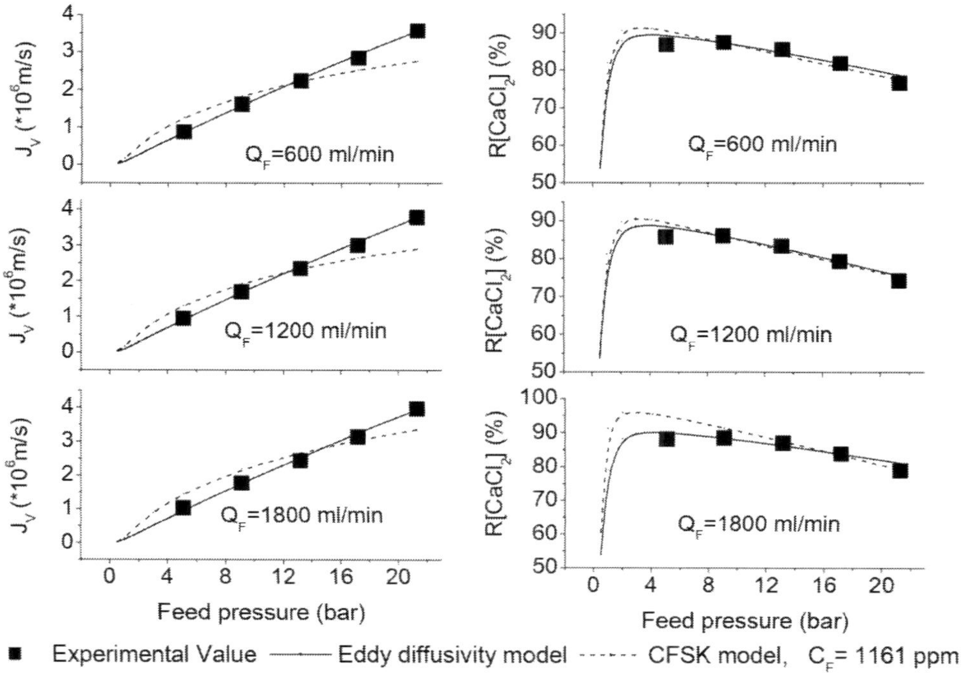

Figure 2: Comparison between CFSK model with EDSK model for $CaCl_2$ –water system at C_F=1161ppm

Table 8: Estimated EDSK and CFSK model parameter of Murthy et al [2] phenol-water system Separation data

C_F	$Pm*10^7$		sigma		$a*10^6$	
	CFSK	EDSK	CFSK	EDSK	CFSK	EDSK
ppm	m/sec	m/sec	(-)	(-)		
13	3.0905	3.0199	0.874279	0.881841	0.50864	0.28835
70	3.6142	3.4140	0.880181	0.882417	0.53585	0.32241

Finally Murthy et al [2] the phenol – water separation data was analyzed by using EDSK model. The estimated parameters are shown on Table 8. The EDSK model predictions of both permeate flux and rejection is found good agreement with Murthy's experimental data as shown in Figure 6 and 7. The average error between the model prediction and measured value of permeate flux and rejection is 12.3 % and .86% respectively. The prediction of EDSK model for phenol water system was compared with CFSK model predictions; the result shows that EDSK model prediction slightly better than CFSK model.

Conclusion:

The rejection of both NaCl and $CaCl_2$ increases and then decreases with increase in feed pressure. The rejection of NaCl and $CaCl_2$ increases and decreases with feed flow rate and feed concentration respectively. The rejection of $CaCl_2$ is higher than the NaCl rejection for same feed conditions. The solute permeability of NaCl and $CaCl_2$ system found to be function of feed concentration. The combination of Spiegler-Kedem and Eddy Diffusivity model predicts much better than the CFSK model. The derived mass transfer coefficient correlation can be used for reverses osmosis system deign and optimization purpose.

Nomenclature:

a,b,c,d	Constant used in correlation for mass transfer coefficient (-)
A	Hydrodynamic permeability ($Kg.m^{-2}.s^{-1}$)
C_b	Concentration of solute at Bulk / Reject stream (kg solute per m^3 of solution)
C_F	Concentration of solute in feed stream (kg solute per m^3 of solution)
C_m	Membrane-solution interface solute concentration at bulk side (kg solute per m^3 of solution)
C_p	Concentration of solute in Permeate (kg solute per m^3 of solution)
F	Parameter used in Spiegler-Kedem model ()
J_v	Permeate flux ($m^3.m^{-2}.sec^{-1}$)
K	Mass transfer coefficient ($m.sec^{-1}$)
M_w	Molecular weight ($kg.kmol^{-1}$)
P_F	Feed pressure (Pa)
P_m	Solute permeability ($m.sec^{-1}$)
Q_F	Feed flow rate ($ml.sec^{-1}$)
R_T	True rejection of solute in equation 2 (-),
R[i]	Observed rejection of solute i = $(1-C[i]_P/C[i]_F)*100$ (-)
	Universal gas constant in equation (1) (8314 $J.kmol^{-1}.^{\circ}K^{-1}$)

Re	Reynolds number (-)
Sc	Schmidt number (-)
Sh	Sherwood number (-)
T	Temperature (°K)
σ	Reflection coefficient (-)
ϕ	Concentration polarization (-)

References

1. Z.V.P. Murthy and S. K. Gupta (1997), Estimation of mass transfer coefficient using a combined nonlinear membrane transport and film theory model, Desalination, 109 pp39
2. Z. V. P. Murthy and S. K. Gupta (1998), Thin film composite polyamide membrane parameters estimation for phenol – water system by reverse osmosis, Sep. Sci. Technol., 33(16) pp2541
3. W. S. Opong and A. L. Zydney (1991), Diffusive and convective protein transport through asymmetric membranes, AIChE J. 37 pp1497
4. Maria Diná Afonso, Maria Norberta de Pinho (2000), Transport of MgSO4, MgCl2, and Na2SO4 across an amphoteric nanofiltration membrane, Journal of Membrane Science 179 pp137–154,
5. Maria Diná Afonso, Maria Norberta de Pinho (1998), Mass taransfer modelling for salt transport in amphoteric nanofiltration membranes, Ind. Eng. Chem. Res. 37 pp4118-4127
6. V´ıtor Geraldes, Viriato Semião, Maria Norberta de Pinho (2001), Flow and mass transfer modelling of nanofiltration, Journal of Membrane Science 191 pp109–128,
7. David G. Thomas, Forced convection mass transfer in hyperfiltration at high fluxes Ind. Eng. Chem. Fundam., 12 (1973) 396-405
8. Wang Xuesong, Mass transfer and fluidized bed intensification of reverse osmosis, Desalination 62 (1987) 211-220
9. Wang Xuesong (1984), J. Chem. Ind. Eng. (China), 2 pp157-163
10. Vassilis Gekas and Bengt Hallström, Mass transfer in the membrane concentration polarization layer under turbulent cross flow : I. Critical literature review and adaptation of existing sherwood correlations to membrane operations (1987), Journal of Membrane Science 30 pp153-170
11. M. Soltanieh and W. N. Gill (1981), Review of reverse osmosis membranes and transport models, Chemical Engg. Comm., 12 pp279

93a What I Learned from Dr. Seibert

Jose Bravo, Shell Global Solutions (US), PO Box 4327, Houston, TX 77210

Frank Seibert has had an indeleble effect in my personal and professional accomplishments. He provided me with some of my early challenges both scientifically and in leadership. He taught me more about fishing than all TV shows put together. I also have admired his dedication to his profession and his family. And more than anything he has been my friend. I would like to share examples of the many influences he has had on me.

163f Temperature Collocation Method for Design of Complex Distillation Column

Libin Zhang, University of Illinois at Chicago, 851 S. Morgan St., SEO 218, Chicago, IL 60607 and Andreas A. Linninger, Bioengineering, University of Illinois at Chicago, 851 S. Morgan St., Room 218, Chicago, IL 60607.

Mathematical models of complex distillation problems are challenging for two reasons. First, problem size increases as the number of stages approximates infinity close to singular points of the composition space (e.g. saddle, pinch points). Secondly, in column design problems it is not known aprior whether a set of desired specifications has a feasible solution. In case of an infeasible specification, all direct solution methods don't converge. Hence, we have developed a set of collocation based methods specially for its use in column design including simple and complex column configuration. The method proposes variable transformations which enable it to successfully solve columns design problems even near pinch and saddle points with acceptable accuracy. We also present a new algorithm to test the feasibility of a distillation design. The key novelty of our procedure exploits a view of physically meaningful design variable to distinguish between feasible region and unattainable composition range. The method is shown to be effective in constant relative volatility, idea and even non-ideal equilibrium models. The presentation will introduce our new algorithm for the complex column configuration.

The new design method features a quick feasible test, design of ideal and nonideal complex distillation column. When defining the polynomials in terms of stage location, it is impossible to model an infinite column section because the polynomials can not be well behaved as the number of stages increase to infinity. The method based the new method can deal with pinch and sharp as well as sloppy split problem. Our algorithm allows us to compute the column section profiles and also can calculate number of trays in each section and the feed tray for each of the separations task. The new collocation method with the help of the column section model reduced significantly the numbers of equations, no matter what the actual number of trays in the complex column is.

This approach also provides means to for determining the minimum reflux ratio with the floppy and sharp split in complex column configuration. Two cases of the separation distillations design about ideal and nonideal mixtures will be discussed.

The new robust feasibility and design algorithm is a necessary component in an optimal complex column sequencing tool. With a very inexpensive feasibility test as proposed by our algorithm, we come one step closes to a fully automated globally convergent column sequencing procedure. This computational approach will be helpful in designing distillation based solvent recovery systems for the pharmaceutical and special chemical industry. In addition, we see the potential of the proposed method in support of plant-wide heat integration in which many feasible test subproblems need to be solved repeatedly.

163d Control Structure Selection for a Dividing Wall Column

Kwang Il Kim[1], Moonyong Lee[2], and Sunwon Park[1]. (1) Chemical and Biomolecular Engineering, KAIST, Guseong-dong, Yuseoung-gu, Daejeon, 305-701, South Korea, (2) Chemical Engineering and Technologies, Yeungnam University., 214-2 Dae-dong, Gyeongsan, Gyeongbuk, 712-749 Korea, Gyeongsan, 712-749, South Korea

A distillation process plays an important role in separating materials but it consumes enormous amount of energy in chemical processes. As an alternative energy saving process, the petlyuk column was proposed in 1950s. The petlyuk column can reduce energy consumption by thermally integrating two columns but integration of two columns involves the extra capital cost. To solve this problem, Dividing Wall Column (DWC), which is conceptually the same as the petlyuk column, has been developed by installing a vertical wall in the middle part of the column.

However, industry has been hesitant to use DWCs because the DWC has control and operation problems. The reason is that general knowledge on the dynamic behavior of the DWC is still lacking because many studies focused on specific processes and materials. To make general design standard for control structure of the DWC, general knowledge of dynamic properties for the DWC is required.

In this study, the dynamic behaviors of the DWC with the various feed compositions and feed materials are studied. By changing the feed components, which have different relative volatilities and mole fractions in feed, the influences of their distribution coefficients on the dynamics and the separation properties of the DWC are examined. Based on the open loop behaviors, reasonable control structures are proposed to deal with various feed conditions. As a result, the properties of DWC and separation performance can be easily predicted by inspecting the composition and the relative volatilities of feed materials.

Acknowledgement

This work was supported by 2006 Energy•Resource Technology Development Program from KEMCO and Center for Ultramicrochemical Process Systems sponsored by KOSEF.

PREPARATION OF Zr(IV)-LOADED MEMBRANES FOR ADSORPTIVE REMOVAL OF As(V) FROM AQUEOUS SOLUTIONS

Shuai-wen Zou, Yu-ming Zheng, J. Paul Chen, Division of Environmental Science and Engineering, National University of Singapore, Singapore.

Introduction

Arsenic can be introduced into water through natural and anthropogenic sources, leaching and soil erosion [1]. Over the past few decades, arsenic pollution is becoming a serious global environmental problem due to its high mobility and toxicity. The long term effects of arsenic pollution could increase the risk of kidney and skin cancers [2]. Therefore many drinking water standards for arsenic have been limited to 10 ppb including WHO and USA [3]. More strict rules and regulations on arsenic in drinking water lead to a great challenge to conventional treatment technologies.

The most commonly used removal technologies for As(V) from aqueous solutions are co-precipitation, ion-exchange and adsorption [4]. Adsorption process is a promising method for the treatment of trace toxic metals present in water due to the low cost and ease in phase separation. However, a detailed study of the adsorption of arsenic on hydrous zirconium haydroxide has not been reported so far.

In the present study, the adsorption performance of Zr(IV)-PVDF membrane for the removal of As(V) under various conditions was carried out. At the same time, the adsorption mechanisms of As(V) on Zr(IV)-PVDF were examined using Fourier Transform Infrared (FT-IR) Spectroscopy and X-ray Photoelectron Spectroscopy (XPS). The objective of this study is dealing with the development of new Zr(IV) loaded PVDF membrane and evaluating the adsorption performance for removal of As(V).

Experimental

Chemicals

Poly(vinylidene fluoride) (PVDF) with M_w of 180,000 and Zirconium(IV) sulfate were obtained from Aldrich Chemical Company. Dodecyl sulfate sodium salt (SDS), ammonia solution, Polyethylene glycol 4000 (PEG) and N,N-dimethyl formamide (DMF) were purchased form Merck. Disodium hydrogen arsenate heptahydrate from Fluka (Switzerland) were used.

Synthesis of Zirconium hydroxide particles

Aqueous ammonia was added into a zirconium sulfate solutions to pH 7.5-8.0 with constant stirring. After precipitation, the precipitate was collected on a filter and washed with distilled water until free of sulfate ions. The powder was dried in air to a constant weight [5].

Preparation of membranes

Table 1 show that the sample ID of prepared PVDF membranes and their corresponding compositions of casting solution, coagulation solution and Zr particles dosage. After stirring the mixtures of PVDF/PEG/Zr particle/DMF at 80 degree and degassing at room temperature, the homogeneous casting solution were obtained. The porous membrane was prepared by coating the casting solution on a glass plate using a film applicator with a gate thickness of 200 μm, then immediately immersed in a coagulation tank at room temperature. The solidified membrane was then immersed in distilled water for 3 days and the water was changed each day. Finally, the Zr-PVDF membrane was dried in air at room temperature.

Table 1
Composition of casting sultions, coagulation baths and Zr dosage

Membrane ID	Casting solution compositions (wt%)			$Zr(OH)O_{1.5}$:PVDF	Coagulation compositions SDS/Water (w/w)
	PVDF	PEG	DMF		
M0	10	2	88	0	0.1%
M1	10	2	88	0.5	0.1%
M2	10	2	88	1.0	0.1%
M3	10	2	88	1.5	0.1%
M4	10	2	88	2.0	0.1%

Characterization of membranes

The surface morphologies of membranes were imaged on a scanning electron microscopy (SEM; JEOL, JSM-5600V, Japan). The surface crystalline structure of membranes was characterized using an X-ray diffraction technique (XRD-6000, Shimadzu).

Adsorption experiments

In this study, arsenic (V) solution from disodium hydrogen arsenate heptahydrate with various initial concentrations was prepared. All the sorption experiments were conducted by adding a dosage of 1.0 $g{\cdot}L^{-1}$ sorbent into airtight bottles containing arsenic (V) solution. In the isotherm experiments, the solution pH was controlled at 3-4. The airtight bottles were rotated on a rotary shaker with a speed of 200 rpm for 48h. The concentrations of the metal ions were measured by an inductively coupled plasma emission spectrometer (ICP-ES; Perkin-Elmer Optima 3000, U.S.A). All the experiments were controlled at 293 ± 0.5K.

Results and Discussion

Crystallization of PVDF

XRD patterns for $Zr(OH)O_{1.5}$, PVDF membrane and Zr(IV)/PVDF membrane were illustrated in Fig. 1. The PVDF membrane shows three characteristic peaks of PVDF crystals at 2θ of 18.3°, 20.2° and 24.4° which are corresponding to (020), (110) and (021) plane respectively. The three characteristic peaks of PVDF and one characteristic peak of Zr(IV) particle can be found in the Zr(IV)/PVDF composite membrane.

Effects of Zr(IV) content on the morphologies of membranes

Fig. 2 shows the SEM images of the top surface of the membranes with different Zr(IV) particles loading. As can be seen, the surface structure of the membranes are significantly affected by the presence of the Zr(IV) particle. The micrograph of Fig. 2(M0) indicates that the pure PVDF membrane has a dense top surface. But with the increasing content of Zr(IV) particles in the casting solution, bigger surface pore size and higher porosity of the membranes can be found, for example Fig. 5(M4), which the Zr(IV) loading is 200% (w/w) in casting solution.

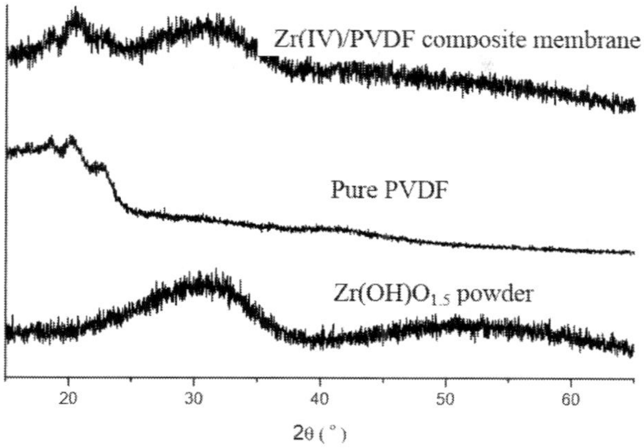

Fig. 1. XRD patterns of Zr(IV)/PVDF composite membrane, pure PVDF membrane and Zr(IV) powder

Effect of pH on arsenic adsorption

Fig 3. shows the effect of pH on the removal performance of As(V) adsorption. The experimental observations demonstrate that more than 95% of the dissolved arsenate was adsorbed by Zr-PVDF in the pH range from 3 to 9. As(V) removal efficiency drops below 40% as the pH increases above a value of 11, which implies the Zr-PVDF can be regenerated using alkaline solutions[6]. The optimal pH range for As(V) removal is 3-6. At pH ranging from 3.0 to 6.5, the predominant ionic species is $H_2AsO_4^-$ with a pK_a value of 2.24, which is liable for the adsorption of As(V) [7].

Kinetic studies

Two batch adsorption experiments with different initial As(V) concentrations were conducted for the adsorption As(V) varying shaking time at constant shaking speed, temperature and initial pH. The adsorption kinetic data of As(V) on Zr-PVDF are shown in figure 4. The adsorption equilibrium is approached in about 40h, which is similar to the reported by Lin and Wu[8], who found that about 40h was needed to reach adsorption equilibrium for the adsorption of As(V) on activated alumina grains.

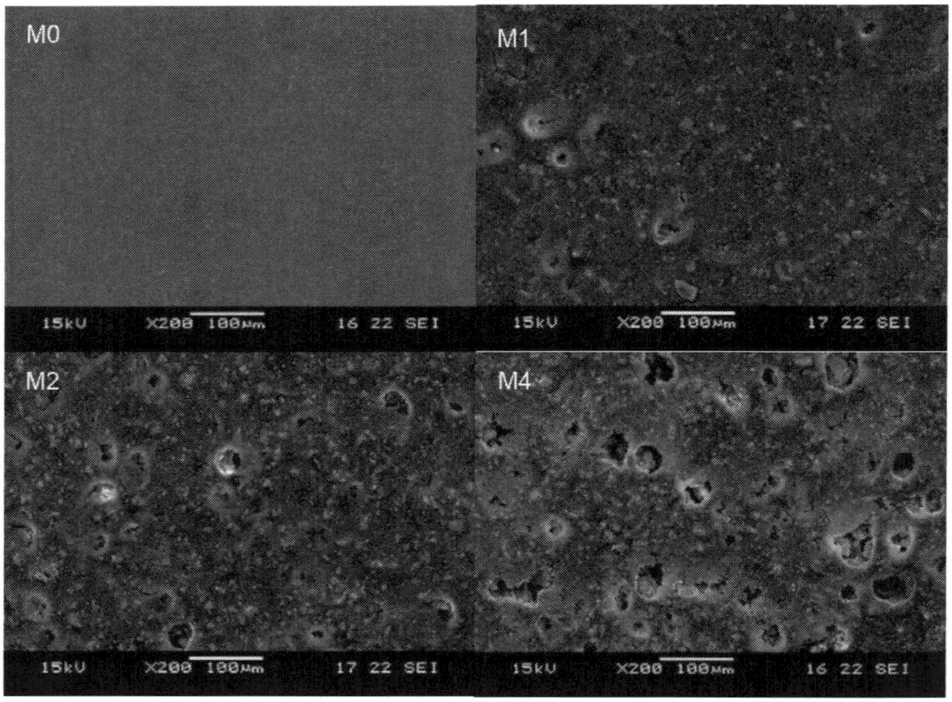

Fig. 2. SEM images of top surface of membranes.

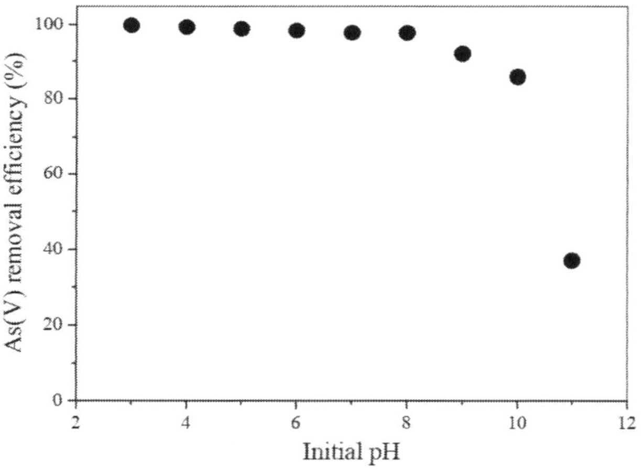

Fig. 3. Effect of pH on the adsorption of arsenate on Zr-PVDF membrane. $[As(V)]_o = 1.0\ mg \cdot L^{-1}$, $m = 1.0\ g \cdot L^{-1}$, $T = 293\ K$, 48h.

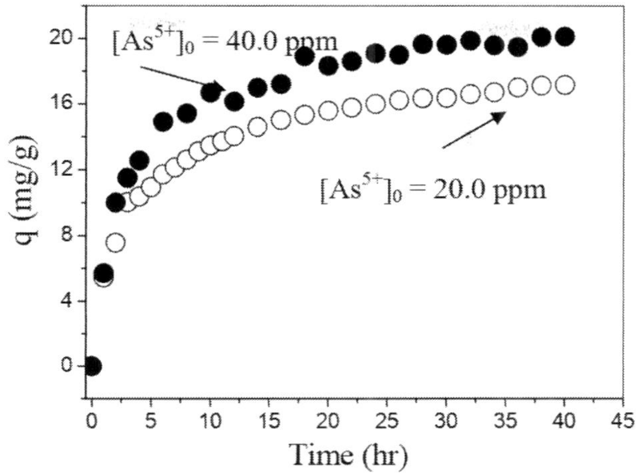

Fig. 4. Adsorption kinetics of As(V) on Zr-PVDF. m= 1.0 g·L^{-1}, T= 293 K, initial pH = 3-4, 48h.

The adsorption kinetic data were further analyzed by pseudo-first-order-rate model and pseudo-second-order rate model. The parameters of the rate models obtained are summarized in table 2. It can be found that the adsorption kinetic data can be well described by the pseudo-second-order rate model with correlation coefficient R^2 of 0.99. In addition, the adsorption capacities calculated by the pseudo-second-order rate model are close to those determined by experiments.

Table 2. Adsorption kinetic parameters obtained by different models

C_0	q_e,exp	First-order kinetic model			Second-order kinetic model		
(mg/L)	(mg/g)	q_{e1} (mg/g)	K_1 (hr^{-1})	R^2	q_{e2} (mg/g)	K_2 (g/(mg.hr)	R^2
20	17.4	10.18	0.0853	0.98	18.35	0.0163	0.99
40	20.2	12.93	0.1057	0.92	21.41	0.0158	0.99

Adsorption Isotherm studies.

Fig. 5 shows adsorption isotherms of the pure PVDF and Zr-PVDF. The As(V) adsorption capacity of pure PVDF is very low, which can be regarded as no adsorption effect. The Zr-PVDF shows a much higher adsorption capacity. The maximum adsorption capacity is 4.5, 10.2, 15.1 and 21.5 mg g^{-1} at the Zr(IV) loading of 0.5, 1.0, 1.5 and 2.0, respectively. It is found the maximum adsorption capacity is 40.0 mg g^{-1} using pure Zr(IV) powder as adsorbent to remove As(V).

The adsorption isotherm data were further fitted by both Langmuir and Freundlich isotherms. The results show that the experimental data are better fitted by the Langmuir isotherm, which indicates that the monolayer adsorption occurs on the Zr(IV) particle surface.

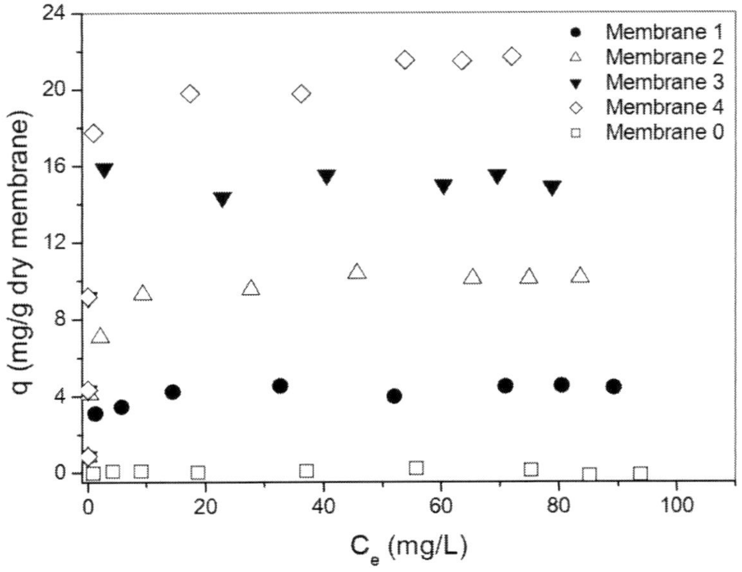

Fig. 5. Adsorption isotherms of As(V) on Zr-PVDF. m= 1.0 g·L^{-1}, T= 293 K, initial pH = 3-4, 48h.

Conclusions

PVDF membranes using Zr(IV) as the additive were fabricated by the phase inversion process. XRD results confirm the existence of Zr(IV) at the membrane surfaces. SEM images illustrate that the increasing content of Zr(IV) leads to an increase in surface pore size and porosity of the membranes. The pH effect studies show that the optimal pH range of adsorption is from 3.0 to 6.0. The results of isotherm experiments demonstrate that data can be well described in terms of the Langmuir equation. The maximum adsorption capacity can be achieved at 21.5 mg g^{-1} of As(V) as the ratio of Zr(IV) particles and PVDF is 2:1. Adsorption kinetics of As(V) at two different concentrations follow pseudo-second-order rate equation.

References

1. A. Brandstetter, E. Lombi, E.E. Wenzel, D.C. Adriano, (2000), "Arsenic contaminated soils: 1. Risk assessment." Dekker, New York, pp. 715

2. M.L. Pierce, C.B. Moore, (1982), "Adsorption of arsenite and arsenate on amorphous iron hydroxide," Water Res., pp. 1247.

3. M. Leist, R.J. Casey, D. Caridi, (2000), "The management of arsenic wastes: problems and prospects." J. Hazard. Mater. B 76(1), pp. 125.

4. M.R. Jekel, in: J.O. Nriagu (Ed.) (1994), "Arsenic in the Environment. Part I: Cycling and Characterization," Wiley, New York, pp. 119.

5. G. P. Shveikin, A. P. Shtin, E.V. Polyakov, T. A. Denisova, Y. N. Blinovskov, I.G. Grigorov, A. N. Khlebnikov, (2005), "Particle-Size Hierarchy in Zirconium Hydroxide Deposits on Inert Supports," Inorganic Materials, 41, pp.557

6. B. Daus, R. Wennrich, H. Weiss, (2004), "Sorption materials for arsenic removal from water: a comparative study." Wat. Res. 38, pp.2948.
7. T. Balaji, T. Yokoyama, H. Matsunaga, (2005), "Adsorption and removal of As(V) and As(III) using Zr-loaded lysine diacetic acid chelating resin," Chemosphere, 59, pp.1169.
8. T.F. Lin, J.K. Wu, (2001), "Adsorption of arsenite and arsenate within activated alumina grains: equilibrium and kinetics." Wat. Res. 35, pp.2049.

200b Feasibility Studies on Dehydration of Ethanol by Distillation Using Salt Effect (on VLE)

Shuzo Ohe, Science University of Tokyo, 1-3 Kagurazaka, Shinjuku-ku, Tokyo, 162-8601, Japan

Ethanol makes an azeotropic mixture with water. Consequently, ordinary distillation processes are unable to purify ethanol. But dissolving salts into the ethanol-water mixture changes the vapor-liquid equilibrium; some salts are able to break the azeotrope. Making use of this salt effect, the azeotropic mixture can be separated into ethanol and water.

We studied the feasibility of dehydrating aqueous ethanol by utilizing the salt effect. For this we found that calcium chloride is the most effective salt. From aqueous solutions of 50 to 75%, simple distillation using the salt effect can recover 93 to 96% of the ethanol. Salt in the ethanol-water mixture must be present at saturation level. After distillation of the ethanol, the salt solution should be further distilled, to remove the water in the still in preparation for the next treatment. Depending on the initial water content, the ethanol recovery is from 88% to 99.9%.

229g Ammonia by Pressure Swing Adsorption – Part II

Felix Jegede[1], James Ritter[2], and Armin D. Ebner[2]. (1) SmartKoncept Technology, 11999 Katy Frwy,, Houston,, TX 77079, (2) Department of Chemical Engineering, University of South Carolina, Columbia, SC 29208

Pressure swing adsorption (PSA) technology enables the energy-efficient recovery of specific compounds from a gas under pressure. At high pressure, gas molecules are preferentially adsorbed onto the adsorbent material depending on their molecular characteristics and affinity for the adsorbent; when the pressure is lowered, the adsorbed molecules are released (desorbed). PSA is currently used for the production of relatively pure oxygen or nitrogen from air and other applications in the petrochemical and gas industries. In this project, sponsored by the DOE, we are developing an ammonia production process that incorporates PSA for recovery of ammonia from the product gas and purification of the process reactants, nitrogen (from air) and hydrogen (from steam methane reforming, SMR). The process converts methane, air, and water to ammonia, with carbon dioxide and a small amount of argon as byproducts. Current ammonia processes react air and syngas (gas mixture of carbon dioxide and hydrogen) to form ammonia, which is recovered via energy-intensive refrigeration and condensation. The new approach will increase ammonia yield and reduce natural gas feedstock and fuel consumption. Improved process heat integration will further reduce energy consumption of the new process and enable generation of excess steam. The improved ammonia process will find application in the U.S. chemical industry, which produced 23.7 billion pounds of ammonia in 2004. Ammonia is also used at power plants in selective catalytic reduction (SCR) and selective non-catalytic reduction (SNCR) systems, which reduce NOx emissions. The new process is scaleable and enables on-demand ammonia production. This will eliminate the need to transport and store large amounts of ammonia, which is classified as a hazardous chemical by the Occupational Safety and Health Administration, at the power plant. Overall, the improved ammonia process utilizing PSA will help reduce the energy consumption and costs in the chemical, petrochemical and power industries.

229a Factors Affecting Overall Tray Efficiency

Henry Z. Kister, Fluor Corporation, 3 Polaris Way, Aliso Viejo, CA 92698

Despite decades of work on tray efficiency and the variables affecting it, engineers are still not clear what is fact and what is myth. Many of the efficiency studies have been contradictory, others have focused on observations derived from laboratory-scale towers that do not fully extend to commercial towers. With much experience gathered in the industry over the years, and a large volume of commercial-scale data released by FRI, there is enough information in the published literature to obtain a closer evaluation of the factors that affect overall tray efficiency. This paper reviews these factors and the extent to which they affect commercial-scale fractionators. It was found that errors in VLE and flow path length often have a significant effect on tray efficiency. On the other hand, factors such as weir height that intuitively are expected to affect overall tray efficiency are only minor contributors. It is hoped that this paper will provide engineers with a guide to the factors they should focus on when revamping a commercial-scale tower or designing a new one.

AUTHOR INDEX